Smart Computing with Open Source Platforms

Smart Computing with Open Source Platforms

Amartya Mukherjee and Nilanjan Dey

CRC Press
Taylor & Francis Group
Boca Raton London New York

CRC Press is an imprint of the
Taylor & Francis Group, an **informa** business

CRC Press
Taylor & Francis Group
52 Vanderbilt Avenue,
New York, NY 10017

© 2019 by Taylor & Francis Group, LLC
CRC Press is an imprint of Taylor & Francis Group, an Informa business

No claim to original U.S. Government works

Printed on acid-free paper

International Standard Book Number-13: 978-0-8153-5952-4 (Paperback)
International Standard Book Number-13: 978-0-8153-5955-5 (Hardback)

Visit the Taylor & Francis Web site at
http://www.taylorandfrancis.com

and the CRC Press Web site at
http://www.crcpress.com

Contents

Preface

The advancement of high-end computation platform, the open source hardware, and software technology is pretty much relevant to make the cutting-edge research in the field of embedded systems, sensor networks, Internet of Things, ubiquitous computing, and machine learning. The sophisticated sensors and hardware devices make the data acquisition task highly efficient for a dynamic and changing environment. The data generated by the ambience get stored in various persistent storage and cloud environments. Further, the data may be processed using highly optimized machine learning algorithms to observe the trends and to predict the future profile of certain environments.

The fundamental objective of this book involves a representation of highly sophisticated open source embedded and modern programming platforms to grab, store, visualize, and predict the data. The main focus of this book is to present an Arduino platform to implement a fundamental solution at a hardware level. Further, the Python programming language is discussed, and the power of an Arduino–Python interface is addressed. Further, machine learning algorithm is introduced, which reflects the practical scenario of Python machine learning and deep learning applications. This book is designed in such a way that readers who are not much familiar with the concepts of Arduino, Python, and machine learning applications can understand and design the applications related to it. This book serves as a guide for undergraduate, postgraduate, and research students, and hobbyists.

Amartya Mukherjee

Nilanjan Dey

Acknowledgment

This book itself is an acknowledgment for various technical and modern scientific computations. We are thankful to the technologies and open source communities, whose continuous support makes this book possible. We are also thankful to our wives, and children for their continuous encouragement.

In real open source, you have the right to control your own destiny.

Linus Benedict Torvalds

Authors

Amartya Mukherjee is an assistant professor at the Institute of Engineering and Management, Salt Lake, Kolkata, India. He holds a bachelor's degree in computer science and engineering from West Bengal University of Technology and a master's degree in computer science and engineering from the National Institute of Technology, Durgapur, West Bengal, India. His primary research interest is in the embedded application development, including mobile ad hoc networking, aerial robotics, and Internet of Things and machine learning. He has written several research articles in the field of wireless networking and embedded systems. His book *Embedded Systems and Robotics with Open-Source Tools* (CRC Press, 2016) is one of the best-selling books in the field of embedded application development.

Nilanjan Dey, PhD, is an assistant professor in the department of information technology at Techno India College of Technology, Kolkata, India. He was an honorary visiting scientist at Global Biomedical Technologies Inc., Roseville, California and an associate member of the University of Reading, London. He has authored/edited more than 40 books with Elsevier, Wiley, CRC Press, Springer, etc., and has published more than 300 research articles. He is the editor-in-chief of the *International Journal of Ambient Computing and Intelligence*, IGI Global. He is the series coeditor of *Springer Tracts in Nature-Inspired Computing* (Springer Nature), *Advances in Ubiquitous Sensing Applications for Healthcare* (Elsevier), and *Intelligent Signal Processing and Data Analysis* (CRC Press). He is an associate editor of *IEEE Access*.

His main research interests include medical imaging, machine learning, data mining, etc. Recently, he is awarded as one among the top ten most published and cited academics in the field of computer science in India during 2015–2017.

Introduction

Smart Computing Methodology

We live in the era of smart devices, where all real-world machines are becoming smarter and smarter. One of the significant events in this context is the introduction of smartphones, smart gadgets, and Internet of Things (IoT)-enabled devices. Data-driven intelligent systems are therefore the backbone of smart systems, perhaps smart environment nowadays. The introduction of intelligent operating systems such as Android makes it highly flexible and versatile to design and deploy smart applications. Also, the enrichment of the embedded hardware and open source rapid prototyping platforms such as Arduino and Raspberry Pi make the life of an intelligent system developer easier. In earlier ages, real-time intelligent systems were highly classified, and they have been used for mission-critical applications such as missile tracking and launching, space missions, unmanned aerial vehicle control and monitoring, and various other areas for military purposes. Currently, numerous applications of smart computing are encountered, starting from data science, data analysis, prediction, and forecasting to smart city and smart home appliances. IoT-based service and comfort management and many other fields of social and commercial applications are supposed to be a much-needed technology enhancement practice indeed.

Smart Computing Platforms

As far as the smart computing platform is concerned, it is perhaps the most challenging platform that takes the input problem, analyzes, and makes a decision that results in a certain level of the solution. There are several smart aspects of smart computing platform that can be considered in this context. Therefore, the primary focus of the computing environment is decision-making. The computing platform may be a software-only platform. Sometimes, it is an independent hardware, but in most of the cases, it is a combination of software and hardware that performs the task in a collaborative manner.

For example, earlier when we came off from the home for a long trip, before leaving our home, we had to check whether the air conditioner, refrigerator, and all other home appliances were turned off properly or not. If we had not checked properly, there would be a high chance of damage of appliances as well as a huge amount of electric bill at the end of the month. Nowadays, a simple solution to this problem is available at the introduction of smart computing and IoT environment. As all home appliances get connected with a centralized gateway. The system monitors the activity of the room. If it is found no activity in the room it will immediately send an alert message to the user and in parallel turn off the home appliances by itself. In another scenario, it was pretty much challenging earlier to prevent load shedding for a certain geographical region due to an uneven distribution of electricity, as no demand-side load prediction mechanism was available. But nowadays due to the intelligent load prediction and forecasting, machine learning algorithm makes the things much easier to predict the demand of the power in the near future.

Microlevel small embedded devices to the low-end minicomputer, high-end graphics processing unit (GPU) and tensor processing unit (TPU) (source: https://developers.googleblog.com/2018/07/new-any-edge-tpu-boards.html) and of course the cloud services are supposed to be the most popular platforms for the intelligent and smart computing ecosystems. The advantage of some of the open source platforms is enormous to design, build, and construct intelligent applications. One initiative called GPUOpen provides a comprehensive collection of visual effects, productivity tools, and various other contents having no cost. The codes here are easily shareable and downloadable. MIAOW GPU is another research developed by the University of Wisconsin–Madison. The fundamental objective of the GPU is to obtain high-performance computation with the free and open source set of code bundles.

Open Computing Language (OpenCL) is another free and cross-platform computing environment. The environment has been built upon the parallel programming of various processors found in the smartphones, personal computers, servers, and other handheld devices. OpenCL improves the performance of image processing, computer vision, and gaming applications significantly.

The Khronos Group announces the release of the Vulkan 1.1, a new accompaniment of the OpenGL platform and the specifications of SPIR-V 1.3. Version 1.1 expands the functionality of Vulkan's core with a large community of the developer's requested features, such as subgroup operations, while integrating a wide range of proven extensions from Vulkan 1.0. Khronos will also release full Vulkan 1.1 conformance tests into open source, and AMD, Arm, Imagination, Intel Corporation, NVIDIA, and Qualcomm have implemented conformant Vulkan 1.1 drivers.

Smart Computation for Embedded Systems and Robotics

Embedded computing and automation robotics are potentially the most emerging technologies that accompany smart computing. Most of the sophisticated embedded platforms use some level of intelligent processing methodology, such as advanced medical system and disease detection and prediction system and fault detection in building structures. Modern 3D X-ray and MRI processing need a vast amount of intelligent signal processing algorithms. There is another wide application of the intelligent embedded system addressed in the field of car manufacturing, where in a car the embedded intelligent tags has been placed that can monitor health of the component, temperature of the engine and cabin, and engine vibration are harvested and recorded for further diagnostics of the fault of not only the current cars but also the cars manufactured in future.

In the field of robotics, an ample amount of intelligent applications can be addressed, such as object and obstacle detection and avoidance, precision take-off and landing of the aerial vehicle, and autonomous navigation of the unmanned aircraft system and missile system. In the medical domain, robotic sugary is popular, which is a relevant example of intelligent robotic systems. In space mission, astronauts use robotic devices for the precise movement of the equipment in a zero gravity zone. Another big example of intelligent robotics system is social robots. One of the popular robots in this context is Sophia. The smart computing algorithms of Sophia are designed in such a way that they can follow the faces, sustain eye contacts, and recognize individuals. A sophisticated neural network of the robot can process the speech. It was developed in the base philosophy of chatbots named ELIZA.

Objective of This Book

This book basically emphasizes the theoretical application of the smart computation methodology and the tools that are used in it. It discusses the details of the design, coding, and implementation as well as the deployment scenario of the computing devices and services. The detailed discussion about the software and the application program interface (API) related to the intelligent system development is addressed. This book specifically focuses on the open source computer languages such as Arduino and Python, which are the core tools of today's cutting-edge technology. It is written in an interactive manner and targeted to the undergraduates, postgraduates, research students, and hobbyists.

Motivation

The main motivation of this book is the implementation while learning. The interactive feature is the primary phenomenon of this book. Various intelligent APIs are discussed to develop smart applications in the field of embedded systems and artificial intelligence. Besides, this book gives a comprehensive guide to learn the programming languages such as Python, Arduino, and Processing. The power of open source development is elevated through this book solely.

How to Use This Book

This book serves as an introductory guide to the concept of open source hardware and software. Unlike the theory-based learning approach, this book primarily deals with implementation-based interactive learning. As the core focus of this book is Python and Arduino, it can be utmost helpful to those people who want to learn Python and Arduino in a simpler and parallel way for developing IoT-based projects as well as hardware implementations for machine learning. The things that are discussed in this book must be implemented after study to get the real essence of the technology.

This book is organized as follows: Chapters 1–7 detail the Arduino programming environment and the language. Chapter 8 discusses the fundamental implementation of the processing language to develop Arduino-based smart applications. Chapter 9 addresses various fundamental Arduino projects. Chapters 10–18 describes the core concept of Python language in an interactive manner. Chapter 19 involves the graphical user interface implementation in Python. Chapters 20 and 21 describe the APIs that are related to the intelligent framework. These chapters also discuss the fundamental and advanced machine learning and deep learning APIs.

1

Introduction to Open Source Hardware

1.1 Open Source Hardware Concept

The term open source hardware [1] is very common today. Open source hardware, also known as open hardware, is basically conceptualized as the electronic hardware or computer component built from some design information that can be copyrighted and licensed without a strict restriction of copyright law [2]. Schematic diagrams, components used, documentation, and the logic design are some of the components of the open source hardware design. Like open source software, the open source hardware may have the source code available for the hardware design; in this case, the source code is the blueprint or a design file that may sometimes be a computer-aided design (CAD) file.

The license of open source hardware involves redistribution and modification of designs and documentations and distribution of any modifications. It also never prevents someone from giving away and selling the project and its documentation. Open source hardware and software designs are also subjected to copyright law, but hardware uses some laws that help them to redistribute and be publicly available [3]. Free license is the most common feature of open source hardware. One of the most common licenses is Creative Common Attribution License. The open hardware must be documented along with design files and source code. Besides, the licensing authority must give permission to modify or change the design files and redistribute them so that anyone can access the modified design files. The fundamental goal of the open hardware is to make and remix the object and reproduce it as easily as possible. However, close hardware uses a strict copyright and patent law, which makes there production of objects very difficult.

1.2 Arduino Ecosystem and Its Types

Welcome to the world of Arduino. Before we start thinking how Arduino works, let us think about some practical scenario. Assume that you are sitting in your office and suddenly you remember that you have forgotten to switch off the air condition in your home. As your home is far away from the office, it will take a good amount of time to reach. Under these circumstances, it is obvious that you would think about a system or service that can stop the air condition from your place. It is also a cool idea if you could design that service or device on your own without spending a good amount of money to buy such a service from third-party vendors. Thus, considering such type of needs, the concepts of devices like Arduino come into play. Arduino was made in such a way that anyone with a minimum knowledge of electronics engineering and programming skill can handle it. The fundamental ecosystem of Arduino comprises a hardware and a software suite. The hardware is the bottom layer on which the bootloader and firmware will run. In most cases, the firmware is associated with the third-party library that is present in the second layer. In the third layer, the communication protocol is present. The main job of this layer is to interconnect other devices that are associated with Arduino. The communication may be wireless or wired. In case of wired communication, some inbuilt protocols, such as Universal Serial Bus (USB) and Inter Integrated Circuit (I2C) are used. For wireless communication, we have to add some additional communication device and a proper library in the layer to properly run the device. On the top of all layers, there are applications that are associated with the device. The application layer ensures the representation of data in a proper format, and sometimes, it is responsible for the visualization of data that are coming out from the layer below it.

The hardware abstraction layer of Arduino is also supported by means of a hardware abstraction library that serves as a wrapper around the Arduino API (Application Programming Interface). The abstraction layer is dedicated for better user experience and code readability (Figure 1.1).

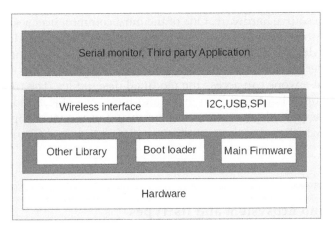

FIGURE 1.1
The layered ecosystem for Arduino application.

1.3 Features of Arduino Hardware

Depending upon its ease of use, Arduino board can be classified into several types [4]. Classification is done based on the microcontroller version and the number of input and output pins, types of microcontroller, Central Processing Unit clock speed, number and type of USB and Universal Asynchronous Receiver Transmitter (UART) interface, size of Electrically erasable programmable ROM (EEPROM), Static RAM (SRAM), and flash memory. There are several types of Arduinos that are manufactured. The most popular Arduino board is Arduino UNO. Figure 1.2 depicted the different versions of Arduino. A comparative study of the different types of Arduino board is listed in Table 1.1.

1.4 Features of Arduino Software

Arduino software package is a set of utilities that primarily have some Integrated Development Environment (IDE) and many standard libraries.

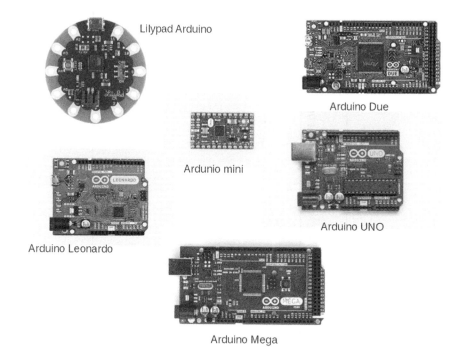

FIGURE 1.2
Different types of Arduino boards.

TABLE 1.1

Comparison of Different Arduino Versions

Board	Microcontroller	CPU Speed (MHz)	Digital I/O	Analog I/O	Voltage (OP/IP)	EEPROM	SRAM	FLASH	USB
Arduino UNO	ATmega328P	16	14/6	6/0	5V/7–12V	1	2	32	Regular USBB
Mega2560	ATmega2560	16	54/15	16/0	5V/7–12V	4	8	256	Regular
DUE	ATSAM3X8E	84	54/12	12/2	3.3 V/7–12V	N/A	96	512	Micro USB (2)
Arduino Pro mini	ATmega328P	8, 16	14/6	6/0	3.3 V/3.35–12V, 5 V/5–12V	0.512, 1	1, 2	16, 32	N/A (UART)
LilyPad USB	ATmega32U4	8	9/4	4/0	3.3 V/3.8–5V	1	2.5	32	Micro
Mega ADK	ATmega2560	16	54/15	16/0	5V / 7–12V	4	8	256	Regular
Leonardo	ATmega32U4	16	20/7	12/0	5 V/7–12V	1	2.5	32	Micro
Ethernet	ATmega328P	16	14/4	6/0	5 V/7–12V	1	2	32	Regular
Mini	ATmega328P	16	14/6	6/0	5 V/7–9V	1	2	32	N/A

IDE made of java requires java to run the whole process. Java runtime environment is by default provided with the suit. The main components of the software are as follows:

- IDE
- Core API
- Core Library
- Sketches
- Driver support for various boards

The Arduino code written in IDE is known as a sketch [5]. The sketch file initially had an extension .pde, which is a common extension for both Arduino and Processing. Later, in the new versions of Arduino, the extension has been changed accordingly; both .ino and .pde are now the supportable extensions. The attached (communication port) COM/Serial port has also been shown in the bottom status bar of Arduino IDE. It also includes the line number of the code. I Arduino IDE—the preference panel—is a good tool to configure the Arduino environment most efficiently. Here, we can not only change the preferred language (system default recommended) and font size but also configure the verbose output for compiling and uploading the code. More additional preferences can be set onto a file called preference.txt.

Whenever we start writing any sketch in Arduino, the environment itself provides access to some core library. It does not require addition of external libraries. For example, the Serial object can be directly accessed through the library Arduino.h which is implicitly available. As Arduino.h is a default library and attached implicitly. We can write the code directly only if we want to make a new library file. We need to add that Arduino.h explicitly in that case. All the libraries are actually written in c++.

There are some API updates that have also been incorporated. pinMode has been updated to support INPUT_PULLUP to add a specific clean support to buttons and switches that are by default high. So, instead of writing pinMode(5,INPUT), we can now write as pinMode(5,INPUT_PULLUP). It gives benefit to set a default value. Universal integer or uint_8, which is a standard 8-bit integer for cross-platform application development, is now supported by most of the functions.

Arduino core API is fundamentally built over Arduino.h header file. If you are in the process of library development, it is necessary to know about the library that has already been included in Arduino.h to avoid including the library. Various header files of Arduino are listed as follows:

```
#include <avr/interrupt.h>
#include <avr/pgmspace.h>
#include <avr/io.h>
#include <math.h>
```

```
#include <stdlib.h>
#include <string.h>
#include  "binary.h"

#include "WCharacter.h"
#include "pins_arduin.h"
#include "HardwareSerial.h"
#include "WString.h"
```

The main program for Arduino is main.cpp. This file always includes Arduino.h., which invocates the void   setup() and void   loop() functions.

The default implementation of main.cpp is shown as follows:

```
#include <Arduino.h>
int main(void){

  init();
#if defined(USBCON)
            USBDevice.attach();
#endif
  setup();

  for(;;)
{    loop();
 if(SerialEventRun) SerialEventRun();
}
return 0;
}
```

In the new version of Arduino software, a new Printable class has been created, and a Print class has been modified. This will directly impact the Stream and Client classes. The new functions have been added to the Print class for more functionalities. Storing a string in flash memory is much easier by virtue of updated string library. We can also use F() command to do so.

1.5 How to Set Up

To set up Arduino, the first step is to download the package from the Internet and install it. At the Arduino website (https://www.arduino.cc/), the following versions are available:

- 32-bit Windows
- 32- and 64-bit Linux

- MAC OSX
- Linux ARM version

We can download the installer or .zip file for Windows and the tar.xz format for Linux. For Windows installation, you have to install the driver file like Arduino. If you first install the Arduino IDE and then connect it with the Arduino board, it will never detect because of the driver issue. Thus, it is necessary to install the driver before you connect the Arduino board with Arduino IDE.

To do so, we have to follow the following steps (Figure 1.3).

We first connect the Arduino board with the computer and go to the Device Manager of the computer. As it detects Arduino, it will immediately show the USB port with the issue (Figure 1.4).

Just right click the USB port. It will immediately show the tab to update the driver. If you click it, you will get the Driver Management window (Figure 1.5).

In the Driver Management window, select Browse my computer for driver software, and it will open the Driver path dialog box. Then, select the location of the driver in the Arduino software package and click Next. Now, it will show the updating driver software (with some force upgrade option,

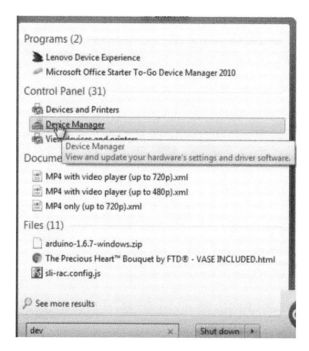

FIGURE 1.3
Windows Device Manager.

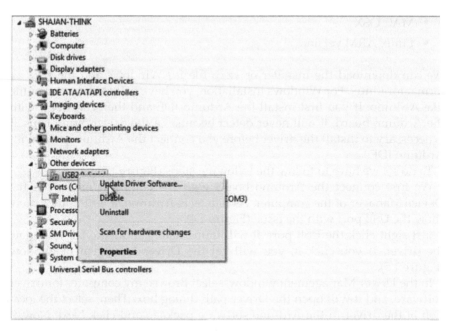

FIGURE 1.4
Device driver installation.

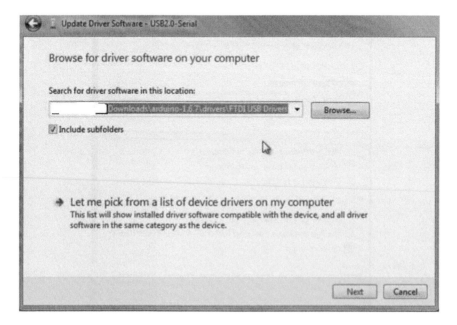

FIGURE 1.5
Driver software browser.

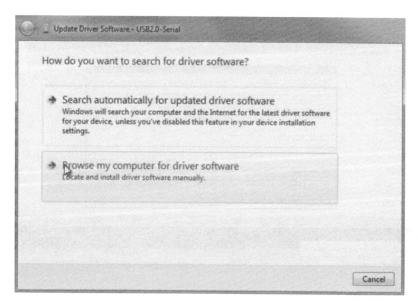

FIGURE 1.6
Installation of the device driver.

just click Yes). After installation of the driver, the USB port will show the COM port.

In case of Linux, generally, it will automatically detect /dev/ttyACM port or /dev/ttyUSB port. This is by default the USB port available in the Linux environment (Figure 1.6).

After the installation of the drivers, we can code and upload the program into the board. During the first-time connection, we first select the board type and the COM/ttyUSB port available in the computer. Your system is now ready to interact with the Arduino board.

1.6 Arduino DUE Environment

Arduino DUE is a special type of microcontroller board. The processor used in this board is an Advanced RISC machine based Atmel SAM3X8E ARM Cortex-M3 CPU.

This Arduino board is perhaps the first Arduino that uses 32-bit ARM CPU. To run an Arduino DUE, we should use the Arduino IDE or web editor. To connect this board with the computer, we should use the USB port that is near the 9-V power connector. After connecting the board with the computer, we should open IDE. Generally, the IDE initially doesn't show the Arduino DUE board in the Board type menu. Therefore, we have to install the library on the

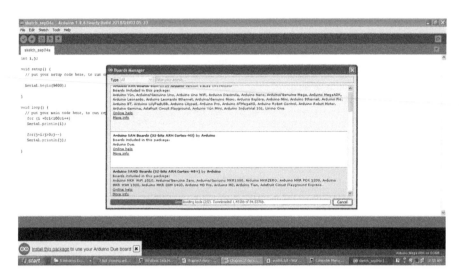

FIGURE 1.7
Arduino DUE installation through board manager.

Arduino DUE board. To do that, we have to open the Board manager shown in Figure 1.7. Then, we have to install the Arduino SAM board Essentials so that the IDE starts supporting ARM cortex-based Arduino such as Arduino DUE. It takes several minutes to install the library and utilities of the Arduino DUE. After installation, the IDE is now ready for running the DUE board.

1.7 Arduino Development Community and Social Coding

As Arduino is an open source platform, it supports the community-based development. The community of Arduino is very large, and it involves students, researchers, hobbyists, technical designers, and many more. The community support ensures free flow of knowledge for the benefit of the community to solve the problem. In a social coding paradigm, any code that you develop is considered to be the contribution to the work of coder community who also will assist you to improve their as well as your own project.

1.8 Concepts Covered in This Chapter

- Arduino Hardware Basics
- Arduino Ecosystem

- Arduino Software
- Arduino Installation
- Development and Social Coding

References

1. Acosta, Roberto. "Open Source Hardware." *PhD Dissertation*, Massachusetts Institute of Technology, Cambridge, MA, 2009.
2. Bonvoisin, Jeremy, Robert Mies, Jean-Francois Boujut, and Rainer Stark. What Is the "Source" of Open Source Hardware? Retrieved from https://openhardware.metajnl.com/articles/10.5334/joh.7/, 2017.
3. Viseur, Robert. From Open Source Software to Open Source Hardware. Retrieved from https://link.springer.com/chapter/10.1007/978-3-642-33442-9_23, 2012.
4. Dey, Nilanjan, and Amartya Mukherjee. *Embedded Systems and Robotics with Open Source Tools*. CRC Press, Boca Raton, FL, 2016.
5. Badamasi, Yusuf A. "The working principle of an Arduino." In *Electronics, Computer and Computation (ICECCO), 2014 11th International Conference on*, pp. 1–4. IEEE, 2014.

2

Arduino Hardware

Arduino is widely known for its open source features. Arduino board primarily have microcontroller and an input–output system. Various versions of Arduinos are available in the market. Typically, the most popular versions of Arduino are Arduino UNO and Arduino Mega. Other versions such as Arduino Nano (version 2.X and 3), Arduino MINI v.04, and Lilypad Arduino are quite popular in the market as well. Some of the latest models available are Arduino Leonardo and Arduino Duo.

This single-board microcontroller based computing device is the best choice for students of various schools and universities, researchers, and hobbyists. The interface circuits, switches and control motor, and various output devices can be done in an easiest way with Arduino. Arduino UNO is a board that is based on the 8-bit microcontroller system known as ATmega328 microcontroller that operates at 5 V [1]. Microcontroller itself has 2 kB of Random Access Memory and 32 kB of internal flash memory.

2.1 System Architecture

The printed circuit board PCB design of Arduino UNO [2] uses surface-mounted device (SMD) components [3]. Each integrated circuit utilized a standard package. Most of the parts are generic and have different functionalities. For example, SOT223 component consists of a transistor or a regulator. The following list depicted the fundamental components of Arduino hardware:

Part	Package Name
NCP1117, NCV1117 regulator	SOT223
LP2985 150-mA low-noise low-dropout regulator	SOT753/SOT23-5
M7 diode system	SMB
LMV358 dual low-voltage rail-to-rail output operational amplifier	MSOP08
FDN340P single P-channel MOSFET	SOT23
ATmega16U2-MU	556-ATMEGA16U2-MU

As Arduino UNO is programmable through USB and, since the microcontroller has no USB transceiver support, the hardware needs a bridge between the serial interface [universal asynchronous receiver transmitter (UART)] and the microcontroller data bus to upload the data successfully. ATmega16U2 is a bridge that has a USB transceiver as well as a UART interface.

2.2 Microcontroller

The Alf and Vegard's RISC (AVR) processor microcontroller is the brain of Arduino. This is an 8-bit microcontroller that supports Reduced Instruction set (RISC) architecture. The uniqueness and versatility of the microcontroller have set apart this microcontroller from any other controller like 8051 [4]. A standard phenomenon of Harvard Architecture is actually incorporated in this system where the system comprises a separate code and program memory that is better than the Von Neumann architecture where program and data have been stored in a single common memory. The AVR family microcontroller first introduces the external flash memory where the programs are stored. Apart from EPROM and EEPROM (electrically erasable programmable ROM), the advantage of flash is that we can store and erase the data from the flash memory as many times we can do. Most of the AVR controller has an extremely nominal amount of EEPROM. The CPU core is shown in Figure 2.1.

Internally, it comprises an AVR CPU [5], timers, serial interface, and Analog/Digital converter. The RISC machine supports 131 instructions and 32 8-bit general purpose register. The supportable clock speed is 20 MHz, million instructions per second (MIPS) rate is 20 MIPS. All peripherals are controlled by control registers. They also configure the functionality of the control registers. Various classes of AVR microcontrollers are available. The designer of the AVR controller must choose the proper part for the specific microcontroller, which may differ from one class to another.

ATmega328p is an AVR family microcontroller that has an 8-bit device. It means it will be having 8-bit data bus and the internal register can handle 8-bit data parallel. The size of flash memory is 32 kB, and it is used for storing the applications. As it is nonvolatile in nature, the application persists even though we unplug the microcontroller. ATmega having 2 kB of SRAM can store the variables used in the application during runtime. It also has 1 kB of EEPROM that stores the data permanently. A block diagram of an AVR microcontroller is depicted in Figure 2.2.

The package is a DIP-28 configuration in which there are 28 working pins including power and input–output pins. Most of the pins are multifunctional and have the advantage of using the same pin for different purposes based

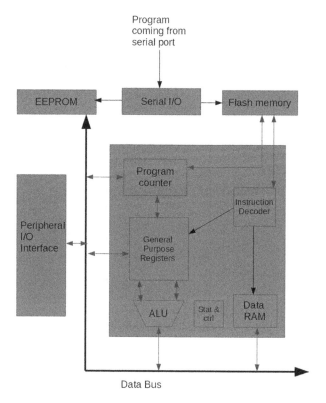

Program
coming from
serial port

FIGURE 2.1
AVR CPU core.

on the configuration that necessarily reduces the pin count of the device. Another popular package is TQFP-32SMD, which is a surface-mounted package that has 32 pins.

The most common feature of the popular microcontroller ATmega32U4 in this family is described as follows:

In the system, the boot program is on-chip, which is often called as bootloader. An 8-bit timer/counter circuit is available for compare and prescaler modes. Also, there are two 16-bit timers/counters for prescaler, compare, and capture modes. It comprises an 8-bit pulse width modulation (PWM) channel with 2- to 16-bit programming resolution. It also has six high-speed PWM channels with 2- to 11-bit resolution. A 12-channel 10-bit analog to digital converter (ADC) is available. The USART of the device has a functionality of Request to Send and Clear to Send) RTS/CTS handshake. Master/Slave serial peripheral interface (SPI) is available in on-chip. A Philips-compatible two-wire serial I²C interface is also available. It also comprises a programmable watchdog timer, an analog comparator, and an on-chip temperature sensor.

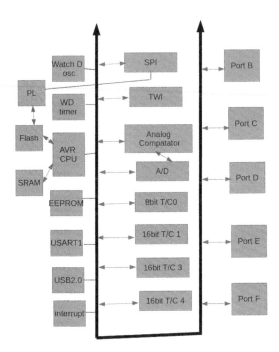

FIGURE 2.2
AVR microcontroller internal architecture.

2.2.1 Power Unit

Arduino accepts a power input of 1.8–5.5 V, but based on the frequency requirement, the voltage range may change. Suppose we want an operating frequency of 20 MHz, the operating voltage should be a minimum of 4.5 V.

2.2.2 Digital I/O of Arduino

Digital I/O of Arduino depends on the I/O port of the microcontroller [6]. This microcontroller unit (MCU) has three ports: PORTC, PORTB, and PORTD. All the pins of these ports can be used as general-purpose digital I/O. Some alternate functions of the pinout are also available, for example, for PORTC, pins 0–5 can be used as ADC input instead of general-purpose I/O.

2.2.3 ADC Inputs

The MCU has a six-channel ADC in PORTC pins 0–5 [7]. The A/D converter has a 10-bit resolution, which means ten different signal levels are used to represent digital data. These pins are directly connected with the headers of analog in Arduino. The analog inputs are not the dedicated pin for the microcontroller. However, they can be used as either a digital I/O or an analog-to-digital converter.

2.2.4 Peripherals in Arduino

UART is a serial interface that is useful for asynchronous data transmission. UNO has only one UART interface [8].

SPI is another interface through which serial transmission takes place. ATmega328 has a single SPI module. SPI's pins from the header next to the MCU in the Arduino UNO board or from the digital header are known as MOSI (pin 11 Master out Slave in), MISO (pin 12 Master in Slave out), and SCK (pin 13 serial clock). MOSI is the master output line to send the data to the slave peripheral, whereas MISO is the slave line that sends data to the master. SCK is useful to produce a clock pulse that synchronizes data transmission generated by the master.

To achieve a maximum speed of the SPI device, the first parameter of SPI setting in the SPI library should be changed.

I2C or two-wire interface is another communication medium in Arduino, where there are two wires called Serial Data (SDA) and Serial Clock (SCL). The last two pins of the digital header of Arduino and pins 4 and 5 of the analog header can be used as an I2C communication line.

In the earlier Arduino model, an additional USB interface chip was present, which is not available in Arduino Leonardo. As Leonardo, Esplora, and micro use ATmega32U4, an additional USB interface is not required, because this has a built-in USB interface. Older models with USB have used Future Technology Devices International (FTDI) chips such as FT232RL. The boards that have no USB in such case were programmed with an external programmer interface.

2.2.5 Detailed Board Configuration

The microcontroller that has been discussed earlier is ATmega328-PU. The headers that have been used in Arduino are IOL and IOH headers. It starts from pin 0 to pin 13, including ground, reference (AREF), SDA, and SCL. The RX and TX pins from the USB are connected to the 0 and 1 pins of Arduino. A 16-MHz ceramic resonator is connected with the MCU via XTAL1 and XTAL2 terminals. A reset pin is provided to prevent unwanted reset in a noisy environment. The pin is connected with a 10 K pull-up resistor. The pin itself has an internal pull-up resistor; however, the AVR hardware design paradigm says that, for the noisy environment, the internal pull-up resistor is insufficient and abruptly resets the device. Reset instruction may come from user side via a button press or through a USB bridge. For high-voltage programming, electrical short, or dielectric (ESD) protection is not given to the Arduino, but it is recommended to add an ESD protection diode from V_{cc} to RESET. Two 100 nF capacitors are attached with the power supply to filter out the noisy supply. These capacitors have a natural property, that is, the impedance of the capacitor decreases with the rate $X_C = 1/2\pi fC$.

2.3 Arduino Interrupts

Interrupt is a procedure to send a message to the processor to immediately hold the current operation and serve a higher priority process. An interrupt can be made forcefully to start a new function (service) by using an interrupt handler. An interrupt handler is a set of instruction or a function that can be called at the point of interrupt. When the function ends, the processor again starts its operation in a normal way.

An interrupt can be generated from several sources such as an interrupt from the Arduino timers, an external interrupt from one of the external pins due to the change of state if the state of a group of pins gets changed then also interrupt occurs.

Arduino has three interrupt timers: Timer0, Timer1, and Timer2. Timer0 is already setup to generate millisecond interrupts to update the millisecond counter reported by the `millis()` function.

Timers are basically simple counters that count at some frequency derived from a 16-MHz system clock. The clock is reconfigurable so that one can alter the frequency at various different counting modes. Timer0 is an 8-bit timer that counts from 0 to 255 and generates an interrupt whenever it overflows. It uses a clock divisor of 64 by default to give us an interrupt rate of 976.5625 Hz (close enough to a 1 kHz for our purpose). We won't mess with the frequency of Timer0, because that would break `millis()`.

2.3.1 Comparison Register

Arduino timers have a number of configuration registers. These can be read or written using special symbols defined in the Arduino integrated development environment (IDE). We'll set up a comparison register for Timer0 (this register is known as OCR0A) to generate another interrupt somewhere in the middle of that count. On every tick, the timer counter is compared with the comparison register, and when they are equal, an interrupt will be generated. The following code will generate a "TIMER0_COMPA" interrupt whenever the counter value passes 0xAF.

2.4 Peripheral Devices and Interfacing

There are various types of devices that can be interfaced with Arduino. The input devices such as keypad and touch screen LCD displays are the fundamental devices that can directly take the input from a human. Other devices such as digital and analog sensors can also be interfaced as an input device.

A UART is a serial interface that is widely used for peripheral interfacing. The ATmega328 has only a single UART module. The pins (RX, TX) of the UART are generally connected to a USB-to-UART converter circuit. It is also connected to pin 0 and pin 1 in the digital header pin. Generally, one can avoid using UART if it has been used as send/receive data over USB.

To interface a keypad, we use a matrix-style keypad interface. Keypad library v3.0 onwards supports multiple key presses. The keypad library is basically based upon the hardware abstraction library. This library is basically a wrapper of Arduino API to enhance code readability and user experience. One of the key parts of the hardware library is Button, and this module is primarily used for button event control and interface. There are many functions and objects that are mainly associated with it.

Button button = Button(12,PULLUP); creates a button object corresponding to pin 12 with pull-up configuration.

Void pullup(); changes the button mode to pull up configuration, and Void pulldown() changes the button mode to pull-down configuration.

boolean isPressed(); checks to see if the button is pressed and returns true if it is. booleanwasPressed(); checks to see if the button was pressed the last time isPressed() was called and returns true if it was.

boolean stateChanged(); checks to see if the state of the button has changed, such as from pressed to released or released to pressed. It returns true if the state has changed since the last isPressed() was called.

boolean uniquePress(); checks to see if the state of the button has changed AND the button is pressed. This will only return true the first time the button was pressed. (As opposed to the isPressed(), it will return true as long as the button is pressed.)

2.4.1 Arduino TFT Interfacing

Thin Film Transistor (TFT) interfacing in Arduino comprises TFT library. This helps Arduino to interface a TFT LCD screen in an efficient manner. It simplified the process of representing the graphical system like drawing different shapes, lines, and images and text to the screen.

The Arduino TFT library extends the property of Adafruit GFX, and Adafruit ST7735 is the main building block. The graphics (GFX) library is solely responsible for performing the drawing task, while the ST7735 library is specific to the screen on the Arduino TFT. The Arduino-specific additions were designed to work similar to the processing API as possible.

Onboard, the screen sometime might have an SD card slot, which can be used through the SD library as well.

The TFT library also relies on the SPI library for communication with the screen as well as SD card and needs to be included in all sketches.

2.4.2 Using the Library

The screen can be configured for use in two different ways. One is to use an Arduino's hardware SPI. The other method is to declare all the pins one by one manually. There is no difference in the functionality of the screen between the two methods, but using the hardware SPI method is significantly faster.

If someone plans to use the SD card on the TFT module, they must need hardware SPI. The entire example library is written for hardware SPI use.

If using hardware SPI with the Uno, you only need to declare the CS, DC, and RESET pins, as MOSI (pin 11) and Serial Clock (SCLK) (pin 13) are already defined.

```
#define CS     10
#define DC     9
#define RESET  8

TFT myScreen = TFT(CS, DC, RESET);
```

To use hardware SPI with the Leonardo, you declare the pins as follows:

```
#define CS     7
#define DC     0
#define RESET  1

TFT myScreen = TFT(CS, DC, RESET);
```

When not using hardware SPI, you can use any available pins, but you must declare the MOSI and SCLK pins in addition to CD, DC, and RESET.

```
#define SCLK 4
#define MOSI 5
#define CS     6
#define DC     7
#define RESET 8

TFT myScreen = TFT(CS, DC, MOSI, SCLK, RESET);
```

2.4.3 Using the Arduino Esplora and the TFT Library

Arduino Esplora is a brand new addition of the Arduino project. It is packed with inbuilt sensors and features. Its cost is around $54, and it looks like a video game controller but is actually more than that. It has temperature, light, and sound sensor that run through an analog multiplexer. With this, it also has a three-axis accelerometer. This is basically based on Arduino Leonardo.

The Arduino Esplora has a socket designed for TFT, and the pins for using the screen are fixed. An Esplora-only object is created when targeting sketches for that board. You can reference the screen attached to an Esplora through EsploraTFT.

2.4.4 Similarities in Processing

Processing is an open source software environment used by designers, artists, and students. The main output of processing is a graphics window on a computer or browser. The Arduino TFT library has made the calls for drawing primitives and text to the screen as "Processing-like" as possible to ensure a smooth transition between the two environments.

2.5 Concepts Covered in This Chapter

- Arduino hardware architecture
- AVR microcontroller
- I/O mechanisms
- I/O interrupts
- Interfacing peripheral module
- Library support for peripheral

References

1. Arduino. *What is Arduino?* Retrieved from https://www.arduino.cc/en/Guide/Introduction.
2. Evans, Brian. *Beginning Arduino Programming.* Apress, Berkeley, CA, 2011.
3. Skyfil Labs. Retrieved from https://www.skyfilabs.com/.
4. Arduino. Retrieved from https://www.arduino.cc/.
5. Faludi, Robert. *Building Wireless Sensor Networks: With ZigBee, XBee, Arduino, and Processing.* O'Reilly Media, Inc., Sebastopol, CA, 2010.
6. Hamblen, James O., and Gijsbert M. E. Van Bekkum. "An embedded systems laboratory to support rapid prototyping of robotics and the internet of things." *IEEE Transactions on Education* 56, no. 1 (2013): 121–128.
7. element14. Retrieved from https://www.element14.com/.
8. Monk, Simon. *Programming Arduino: Getting Started with Sketches.* McGraw-Hill Education TAB, New York, 2016.

3

Data Types, Operators, and Expressions

There are various primitive data types that have been supported by Arduino programming environment. Majority of the primitive data types, in this case, have similarity with other standard programming languages such as C and C++ [1]. Arduino supports user-defined data types such as structure and enumeration. Pointers declaration is also possible in the Arduino environment. They are special types of variables in which the addresses of the data elements are stored. Along with primitive types, some of the special data types are also present in an Arduino environment such as string and word. These types are mostly used for storing a bunch of characters at a time. Data types such as Boolean also exist in this environment.

3.1 Primitive Data Types

3.1.1 Integer

Integer is the data type used to store whole numbers [2]. In an Arduino environment, it takes 2 bytes (16 bits) to store integer data. In a standard ATmega-based Arduino UNO, the range is −32,768 to +32,767. In case of Arduino Due and SAMD-based (Software as a Medical Device) boards, the size of an integer is 32 bits, and hence, the range will increase by -2^{31} to $+2^{31}-1$. Normally the integer data are signed. It means that it may store either a positive or a negative value. Normally, it uses a radix complement mathematics (2's complement in this case) to store and represent integer data. For a 32-bit integer data, the most significant bit (MSB) can be used as a flag to represent the sign of the number, often called sign bit. The remaining part can be used to represent the value.

In most cases, integer data types are used to create a data variable for storing pin numbers or maybe sometimes sensor data in a raw integer format. The example is given as follows:

```
int ledPin = 12;
```

3.1.2 Byte

Another form of an integer is a byte. It is an 8-bit-wide unsigned data element. The maximum allowable value is 0–255. Mostly, the declaration of a byte is as follows:

```
byte b= B1010;
```

Here, B stands for a binary formatter that tells the system to take the data element in a binary form. Another use of byte is a byte() that is primarily used for byte conversion or byte casting. This method is used to convert a value to a byte data type.

3.1.3 Other Types of int

The keyword "int" is fundamentally used as an integer data value that must not have a negative number. The unsigned integer has a range of 0–65,535. To make the number unsigned, we commonly use an unsigned qualifier as we do in C or C++ [3] language. Often, we use another version of integer known as long integer. Here, long can be used as a qualifier to increase the size of the integer data. Commonly, it occupies 4 bytes of data. The most popular use of such a qualifier is to declare a time data that count the time in an *unsigned long* format.

   Short is another form of integer data type. It is 16 bits wide for all devices and can be used as a replacement for the integer data sometimes.

   int16_t (or sometimes unit16_t) is another special form of data type declaration used in Arduino; this perfectly specifies that the size of an integer is 16 bits irrespective of any platform (such as Arduino Due). This will ensure the portability of the data, because in Arduino, it is quite possible that the same code may run in various machines. Therefore, the portability should be of high concern in this matter.

3.1.4 Integer Constant

Various forms of an integer constant are allowed in an Arduino sketch. The constants look like normal integer data in a certain format. We can use a decimal number directly or we can apply fora binary, octal, or hexadecimal number as constant. Some of the examples of the constant are shown in Table 3.1

3.1.5 Characters

Characters are the fundamental data type in Arduino. Some of the sensors also supply the input data through character format. Characters are also used for output display purposes. Most of the projects may consist of light

TABLE 3.1

Format for Integer Constant in the Arduino Sketch

Type	Format
Decimal	234
Binary	B1001 (leading B means binary)
Octal	0142 (leading 0 means octal)
Hexadecimal	0X143 (leading 0X means Hex)

emitting diode display (LED) or liquid crystal display (LCD) screen as an output device for Arduino, and thus the character type has a significant role in Arduino sketch.

Practically, character uses 1 byte of memory for storage and can be decoded in the form of 0–255 decimal number, often called as ASCII (American standard code for information interchange) sequence. This convention is applicable for unsigned char. A signed char can hold integer values in the range of –128 to +127.

3.1.6 Floating Point Data

There are two standard data types to store floating point data: float and double. Both the float and double data have the same size of 4 bytes (32 bits). The float data type stores a decimal value of 6–7 digits of precision. To get more amount of precision, we can use the double data type. In the double data type, it generally stores a decimal value of up to 15 digits of precision. The computation of the floating point data elements generally takes more amount of time in comparison with integers; therefore, we should take extra care for the declaration of the variable. Generally, a mixture of integer and floating point expression will produce a floating point result having 0 in all decimal places, which is equivalent to an integer value. Therefore, to get the actual value, we sometimes perform typecasting.

```
int a =5; float b=2.5;

float c= (float) (a/b);
```

The double precision data type takes only 32 bits for Arduino UNO and other ATmega-based systems, whereas Arduino Due takes 8 bytes of memory.

3.1.7 Array and String in Arduino

Like other programming environments, Arduino also supports the implementation of an array. The general syntax is as follows:

```
data type array [size];
```

An array may be of any data type, including some user-defined data types. Generally, arrays are highly useful to store the sequence of sensor data, keypad map, or I/O pin sequences.

One can design a one-dimensional or a two-dimensional array. An array typically maintains a zero-based addressing mechanism, which means the starting element will start from index 0 and the last element will be an index of 1 less than the total number of elements.

Accessing an array can be made using a loop, such as for, while, or do-while (which will be discussed later). The syntax for the same is as follows:

```
int sensorData[100];
int SIZE =100;
for (int i=0 ;i<SIZE ; ++i)
sensorData[i]= Serial.read();   // store data that
are available in serial port.
```

An array can also store a set of characters known as a character array. It can be declared in the following manner:

```
char dataStream[200];
```

A two-dimensional array can also be created in Arduino. It is mostly useful to create an Input map of a keypad or output display data grid. Also, it is used to realize the sensor map for sensor networks as well.

The definition for the two-dimensional array is shown as follows:

```
int Map[100][100];
```

The accessing mechanism of the array is also similar to the C language. The accessing mechanism of a two-dimensional array is given as follows:

```
int keymap[10][10]; int k =1;

for( int i=0;j<10;i++) {
   for(int j=0;j<10;j++){
      keymap[i][j] = k;
       k++;
      }
  }
```

3.1.8 Strings

Like other languages in Arduino, strings are nothing but a set of characters terminated with a null [4]. This null (\0) terminator (ASCII value 0) tells the Serial.print() function to terminate the string. Otherwise, it may read the subsequent bytes in the memory which are not the actual part of the string.

A long string can be a wrap by applying leading and trailing double quote ("") symbol if the size of the string is too large and takes more than two lines to represent.

```
char string[] = "hello"
                " this is"
                " test";
```

Here, the compiler takes the characters of a string within "" and merges them accordingly until it reaches the terminator (;).

Practically, in most of the cases, we use a text message to display on the screen for different applications. In such cases, we often use an array of strings, which is physically a two-dimensional array of the characters followed by null in each row. In this case, we can declare the strings in the following way:

```
char string[] ={"string1","string2","string3",String4"};

void setup(){
    Serial.begin(9600);

    }
void loop(){
    for (int i=0;i<4;i++){
        Serial.println(string[i]);
        delay(1000);
    }
}
```

Often, we use to transmit the string data from one device to another device in many Arduino projects. For example, if we want to transmit a string set through a 433-MHz RF (radio frequency) module, we preferably use a function called sprintf() function. This function prints data elements of any type into a string buffer that is sent through the wireless communication device.

The example of sprintf() function is given as follows:

```
unsigned int i = 0;

                            // unsigned integer declaration
void setup() {
Serial.begin(9600);
}

void loop() {
char buffer[50];

    int a = 10, b= 20, c;
```

```
c = a + b;

sprintf(buffer, "Sum of %d and  %d is %d", a, b, c);


Serial.println(buffer);
 }
```

In this case, the string "Sum of 10 and 20 is 30" will be stored in the buffer. Further, it will get printed in a serial monitor or may be sent through any communication medium.

3.2 User-Defined Data Types

Arduino language supports user-defined data types. Most popularly, structure type and Enum types are used.

The structure type involves keyword struct that binds all other primitive data types altogether. Enum is another type that consists of a set of constants, and the value of the variable should remain within the boundary of that constant.

The structure can be defined as a group of the primitive data type. It should start with struct followed by the name of the structure.

```
struct RGBcolor{
                byte R; // members
                byte G;
                byte B;
                };
```

We can initialize the data member in the setup() function in the following way:

```
void setup(){
                struct RGBcolor   color;
                color.R = 0;
                color.G = 0;
                color.B = 255;


        if ( color.R == 0 && color.G==0 && color.B=255){
                Serial.print("code matches BLUE");
                }
```

The dot (.) operator is primarily used to access a member of the structure in this case.

In this case, we use enum keyword to create an enumerated data type. The structure of the enumerated data type declaration is shown as follows:

```
enum ternary {  unknown, true, false };

    ternary variable = unknown;

    if (variable==unknown){
      //do code
      variable = (condition ? true : false);
    }
```

3.3 Declaration of a Variable

Variable declaration in Arduino is pretty straightforward. We generally use the following format for declaring a variable:

```
Data type variable_name;
```

Such a declaration is known as a static declaration. The memory will be allocated during compile time. We can make a dynamic declaration of the memory using dynamic memory allocation. We can use `malloc()` function to do that.

```
uint8_t  x = (uint8_t)malloc(sizeof(x));
```

Here, `malloc()` allocates the memory based on the size of the data type. In Arduino, along with malloc, we could use the `free()` function. This function deallocates the memory created by malloc. In Arduino, we must take care of dynamic memory allocation because the size of the memory is very limited.

One of the thumb rules you can apply while declaring a variable is the reduced oversized variable. Use the variable as per your requirement. Do not use any oversized variable. For example, `int` takes 2 bytes, but sometimes we can store the data of 1 byte in an int variable. In such a case, it is better to use `uint8 _ t` that occupies 1 byte of memory. For larger values, it is better to use `int`, `word`, or `uint16 _ t` [5]. For even more larger data, we can go for a 4-byte data type such as `long` and `uint32 _ t`. When a variable is declared, it occupies memory in SRAM. In a parallel system, stack and heap are also allocated. Global variables directly occupy SRAM memory, and they push the heap towards stack. As a result, memory overflow occurs.

3.4 Operators

In Arduino, numerous operators used are quite similar to the C and C++ languages. In various application designs, the commonly used operators are arithmetic operators, logical operators, and bitwise and shift operators.

Operators such as +, −, *, /, and % are fundamentally used for the arithmetic operation. The + operator cannot be used as for concatenation. Logical operators such as &&, ||, and == are used for logical comparison. The ! operator, which is often used for negation, is a unary operator. Some of the examples are shown as follows:

```
int a=0,b=7,c;

c= a&&b;   // c will be 0 since one of the value amongst a and
b is zero.
c = a|| b; // produce c =1 since one of the value is nonzero.
```

The == operator checks the equality between two or more variables.

Bitwise operators are used to perform bitwise operations on bit patterns. Some of the bitwise operators are discussed as follows:

- Bitwise AND (&) performs ANDing operation between the bit patterns of the data provided on the left- and right-hand sides of the operator.
- Bitwise OR (|) performs bitwise OR of two numbers provided on the left- and right-hand sides of the operator.
- Bitwise operators such as bitwise XOR (^) and bitwise not (~) are used to perform exclusive OR and not respectively.

The example is given as follows:

```
byte a = 2,b=3,c;
c= a&b;     // performs 00000010&00000011 results 00000010
c=a|b;      // performs 00000010 | 00000011 results 00000011
c=a^b;      // performs 00000010&00000011 results 00000001
```

There are two shift operators that can be widely used in Arduino: right shift >> and left shift <<. The operation procedure of the left and right shifts is similar to C++.

The format for the same is shown as follows:

```
value = data >> number of shift;
value = data << number of shift;
```

3.5 Precedence and Associativity

Precedence is a rule through which a compiler can judge the priority of an operation. Generally, multiplication and division operators have a higher precedence compared with addition and subtraction operators. The parenthesis operation has the highest precedence.

The associativity of the operator is only applicable when two or more operators having the same precedence are encountered in an expression. In that case, the expression may be evaluated either from right to left or from left to right. Table 3.2 shows the precedence and associativity of various operators used in Arduino.

TABLE 3.2

Precedence Table

Operators	Description	Associativity
( )	Parenthesis	Left to right
.	dot	
→	Arrow	
++, −−	postincrement/postdecrement	
++, −−	Preincrement/predecrement	Right to left
+, −	Unary +, −	
!, ~	Unary not, bitwise not	
(type_cast)	Typecasting	
*	Value at address	
&	Address of	
sizeof	Size of	
*, /, %	Multiplication, division, modulo	Left to right
+, −	Binary +, −	
<<, >>	Logical operators	
< >=		
== !=		
&	Bitwise and logical AND, OR, and	Left to right
|	bitwise XOR	
^		
&&		
||		
? :	Ternary and all assignment	Right to left
+= −=	operators	
*= /=		
>>=		
^= |=		
&= %=		
=		
,	Separator	Left to right

3.6 Concepts Covered in This Chapter

- Allowable data types
- Variable declaration
- Various operators
- Precedence and associativity rules

References

1. engineering.com. Retrieved from https://www.engineering.com/.
2. Monk, Simon. *Programming Arduino Next Steps: Going Further with Sketches.* McGraw-Hill, New York, 2014.
3. Barton, John J., and Lee R. Nackman. *Scientific and Engineering C++: An Introduction with Advanced Techniques and Examples.* Addison-Wesley, Boston, MA, 1994.
4. Brock, Dean J., Rebecca F. Bruce, and Susan L. Reiser. "Using Arduino for introductory programming courses." *Journal of Computing Sciences in Colleges* 25, no. 2 (2009): 129–130.
5. Olympia Circuits. *Learn Arduino with Olympia Circuits.* Retrieved from http://learn.olympiacircuits.com/.

4

Functions in Arduino

The function-oriented programming approach is the fundamental backbone of the Arduino programming. Each module in the firmware design can be considered as a set of functions. Like other programming languages such as C or C++ [1], we can define a set of functions in a single file or might write them in a separate header file and attach them using the `include` directive. Here, functions are also treated as modules. Unlike other programming languages, the Arduino function structure does not depend on the `main()` function. The modular programming approach is the key benefit of the Arduino firmware design. Fundamentally, we can define various functions within Arduino depending on the maximum limit of memory. When a function is created and compiled, it will become a code segment that will store into the program memory of Arduino (typically SRAM) [2]. When a call is made, the address of the function is fetched as a subroutine. Arduino uses various libraries for various operations. All libraries are mainly populated with numerous kinds of functions that are used for different applications. Some of the great advantages of function in the context of Arduino programming language are listed as follows:

- A function is a tool through which the program can be conceptualized easily and the program looks organized.
- It reduces the chances of error that may occur in a large program segment and hence reduces the robustness of programming.
- It increases the reusability of the sketch, and the size of the code segment becomes smaller.

4.1 `setup()` and `loop()` Functions

`setup()` and `loop()` are the fundamental functions in Arduino [3]. The syntax of the `setup()` function is written as follows:

```
public void setup(){
//body
}
```

It initializes the environment. The entire initialization task has to be written in this function. The `setup()` function can be written as `public void` or sometimes void. The publicvoid is an access modifier that tells that the function is publicly available. The void is the return type that tells that the function by default returns nothing. As the setup() function is an initializer, we can mostly perform variable initializations, serial port initialization, and pin mode setup, or start using any libraries. When sketch starts running, the setup() function executes once during the very beginning of the system initialization.

```
int ledpin;
Void setup() {
        Ledpin = 12; // initialization of pin
        pinMode(ledpin,OUTPUT); // assigning the mode
        Serial.begin(9600); // initialization of the serial
communication baud rate.
 }
```

public void loop() is another important function that is primarily used for performing repetitive tasks. The return type here is also void. The loop() function by default iterates in such a way that the code inside the loop will run in an infinite sequence. This is highly useful for an embedded computer system to grab the data from sensors having a sequence of iteration or might be to print the series of values in an LCD or LED screen. We cannot control the execution of the loop function; however, it can be slowed down using a sleep call inside the loop. The structure of the loop() function is shown as follows:

```
public void loop(){
// code that is repetitively executable
}
```

We can accommodate loops in a loop() function too. Doing so, it will produce a nested loop-like situation. This operation is highly useful to control the actuators of robotic systems. A detailed code implementation is given as follows:

```
void loop(){
        digitalWrite(ledpin,HIGH); // perform write
operation at pin mentioned
        input = analogRead(A0); // read analog data from
analog pin A0 and store it to input
        Serial.println(input,DEC);
    }
```

4.2 User-Defined Functions

In an Arduino programming environment, we can also accommodate some user-defined functions [4]. The user-defined function can also be treated as a library that can be attached as header file with the main source code. The function call can be made from both setup() and loop(). The following example shows a user-defined function:

```
void setup(){
      Serial.begin(9600);
      printstar();
      Serial.println("* Entry menu *");
      printstar();
}
void loop(){
          // do nothing
}
void printstar(){
                int max=20;
          for(int i=0;i<20;i++)
                Serial.print("*");      //print * 20 times
  Serial.println(" "); // print next line
  }
```

The above-mentioned program shows how we can use a user-defined function. In this case, printstar() function is basically used to print 20 * symbol. The program segment shows the structure of the code. We can create a user-defined function anywhere in the program, that is, before setup(), after loop(), or in between setup() and loop(). In this case, typically, it has been written after the loop() function. The printstar() function simply performs the printing of 20 * symbol by applying a loop from 0 to 19, and after doing that, a Serial.println() call will be executed so that it will move to the next line. The important observation is that here the printstar() function has called from the void setup() function; therefore, the function will be called based on the number of calls that have been made from the function. Overall, two calls have been made, so 20 stars were printed twice. Here, the loop() function is a do-nothing function.

4.3 Recursive Function Calls

A recursive function is a function that can be called from the body of the function itself. Like other programming languages, Arduino also supports a recursive call. The following call will produce a recursive call in Arduino:

```
void setup(){
        callr(20);
}
```

```
void loop(){ }
void callr(int n){
                Serial.print(n,DEC);
        If (n==0)
        return 0;
        else return callr(n-1);
    }
```

The above-mentioned example will produce a recursive call up to 20 times depending on the value of *n* at each iteration of call, and the value will be decremented by 1.

The major thing that we should take care in the recursive function is the number of calls that have been created. Since Arduino contains a very nominal size of Random Access Memory (on the order of kilobytes), the number of recursive calls is sacrificed. It is observed that Arduino UNO or Duemilanova can run a recursive call up to 929 times before it freezes. The number of iterations may vary depending on the size of the dynamic stack created by the system within its RAM.

For a genuine recursive program, some calculation and manipulation are required. Also, we must have an approximation of the memory state at the time of the recursive call. For that, we have to visualize the organization of the SRAM of AVR-based Arduino (Figure 4.1).

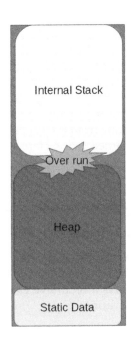

FIGURE 4.1
Arduino memory organization.

Then the total size of SRAM should be computed (depends on Atmel MCU and what kind of Arduino board is used).

On this diagram, it is easy to find out the size of *Static Data* block as it is known at compile time and won't change later on.

The size of heap can be more difficult to know and manage as it may vary at runtime, depending on dynamic memory allocations (malloc or new) performed by your sketch or the libraries. Using dynamic memory is quite rare in Arduino, but some standard functions do it.

The Stack size will also vary during runtime, based on the current depth of function calls (each function call takes 2 bytes on the Stack to store the address of the caller) and the number and size of local variables, including passed arguments (that is also stored on the Stack) for all the functions called until now.

So let's suppose your recurse() function uses 12 bytes for its local variables and arguments, then each call to this function (the first one from an external caller and the recursive ones) will use 12+2 bytes.

If it is assumed that

- We are on Arduino UNO (SRAM = 2 K).
- Our sketch does not use dynamic memory allocation (no Heap).
- We know the size of your Static Data (let's say 132 bytes).
- When your recurse() function is called from your sketch, the current Stack is 128 bytes long.

Then, you are left with $2,048 - 132 - 128 = 1,788$ available bytes on the Stack. The number of recursive calls to your function is thus $1,788/14 = 127$, including the initial call (which is not a recursive one).

As you can see, this is very difficult but not impossible to find what you want.

A simpler way to get the stack size available before recurse() is called would be to use the following function (found on Adafruit learning center; I have not tested it myself):

```
int freeRam ()
{
  extern int __heap_start, *__brkval;
  int v;
  return (int) &v - (__brkval == 0 ? (int) &__heap_start :
(int) __brkval);
}
```

For Arduino, it is better to perform tail recursion rather than a normal recursion because tail recursion physically takes less amount of memory in comparison with normal recursion. As a result, it will produce a good recursive performance bay avoiding the recursive call.

4.4 Library

Arduino supports various libraries to extend the functionality of the environment. Working with data and to manipulate them, various hardware libraries are needed. There are several library functions that are inbuilt along with the default Arduino IDE. To support any third-party devices or different shield, generally, we use the libraries that are compatible with that hardware. The following text describes the list of libraries that are inbuilt in the Arduino. In Arduino IDE, go to Sketch and from that select import library. You will see numerous predefined libraries available in Arduino (Figure 4.2).

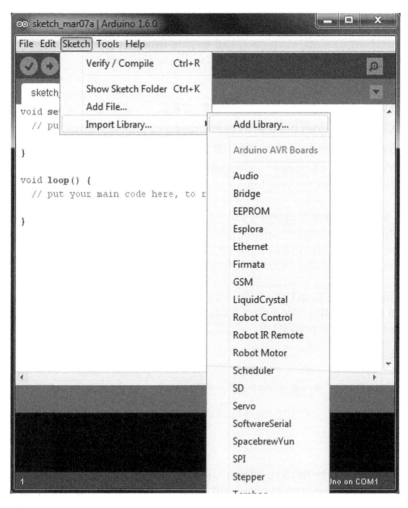

FIGURE 4.2
IDE with the library tab.

4.4.1 EEPROM Library

Arduino contains AVR-based microcontroller in which a small-size EEPROM is present. This memory is basically a nonvolatile kind of memory that can store the data bit even if the power goes down. The microcontroller board has various sizes of EEPROM depending on its configuration, for example, ATmega328 has 1,024 bytes of EEPROM, ATmega168 and ATmega8 have 512 bytes, and ATmega1280 and ATmega2560 have 4 kB of memory.

The fundamental functions that are incorporated into the EEPROM library are described in the following text.

4.4.1.1 *read()*

read() function reads the byte from EEPROM. The syntax is EEPROM. read(address). Where the address is the location supplied in the form of integer, it returns the byte data stored into that location. To use the function of EEPROM library, we have to use the header file EEPROM.h. The following program shows the EEPROM read() function operation.

```
#include <EEPROM.h>

int adr = 0;
int val;

void setup()
{
  Serial.begin(9600);
}

void loop()
{
  val = EEPROM.read(adr);

  Serial.print(adr);
  Serial.print(" >> ");
  Serial.print(val);
  Serial.println();

  adr++;

  if (adr == 512)
    adr = 0;

  delay(1000);
}
```

The above-mentioned program reads the location 0–512 and prints the data from it. As the address goes to 512, it will again reset to 0.

4.4.1.2 *write()*

write() function performs writing data bytes on EEPROM. The syntax is somehow similar to the read() function, that is, EEPROM. write(address,data). The maximum value that we can write in an EEPROM is 255. EEPROM generally takes 3.3 ms to write. An EEPROM memory can be rewritten up to 100,000 times write/erase cycle. A sample code for writing into EEPROM is shown as follows:

```
#include <EEPROM.h>

void setup()
{
  for (int k = 0; k < 255; k++)
    EEPROM.write(k, k);
}
void loop()
{
// do nothing
}
```

4.4.1.3 *put()*

put() function is mostly used to write a data type or an object in EEPROM. The syntax is EEPROM.put(address,data). This function uses the EEPROM. update() function to perform a write() operation, and it does not rewrite the values. The following example shows how put() can store a float and a user-defined data type in an EEPROM.

```
#include <EEPROM.h>

struct Obj1 {
  float f1;
  byte f2;
  char name[5];
};

void setup() {

  Serial.begin(9600);

  float f = 123.456f;
  int eeAdr = 0; //Location of the data.

  EEPROM.put(eeAdr, f);
  Serial.println("float data has been written!");

  /*storing user-defined data type using put */
```

```
Obj1 var = {
    3.14f,
    65,
    "Work"
};
```

```
eeAdr += sizeof(float); //Move address to the next byte
location after float 'f'.
```

```
EEPROM.put(eeAdr, var);
Serial.print("user defined data type written! \n\nrun
eeprom_get to see how you can retrieve the data!");
}
```

```
void loop() {
    // do nothing!
}
```

4.4.1.4 get()

An Arduino board can hold the data even if the power is shut down. To get the data that are stored permanently into Arduino board, we generally use the get() function.

```
#include <EEPROM.h>
```

```
void setup() {

    float f1 = 0.00f;
    int eeAddr = 0;

    Serial.begin(9600);
    while (!Serial) {
        ; // waiting for the connection of serial port
    }
    Serial.print("Read data from EEPROM: ");

    EEPROM.get(eeAdd, f1);
    Serial.println(f1, 3);      //in case the data format not
matches it may print 'ovf or nan'

}
```

```
void loop() {}
}
```

4.4.1.5 update()

update() function only performs when the data bit written in the ROM differs from that of the previous data. The parameters of this function are the address and data value. The update also takes 3.3 ms to perform its operation. The format is mentioned as follows:

```
EEPROM.update(addr,data);
```

4.4.2 Firmata Library

Firmata is a communication protocol that enables the microcontroller to communicate with the computer systems, smartphones, tablets, and so on. It can be implemented as a firmware for the microcontroller of any architecture or maybe the software package for any system platform.

The basic format for Firmata is called midi message format. In this case, the command consists of 8 bits and the data consists of 7 bits. The command midi channel pressure (0XD0) is a 2-byte command, and it is used to enable reporting for a digital port.

Current protocol version for Firmata is Firmata2.6.0. The intention of this protocol is to allow a number of the microcontrollers to be controlled as possible through a host computer. This protocol was fundamentally engineered for the direct communication between a microcontroller device and the software object on a host computer system. The host software object, in this case, must provide a user interface that makes sense in that environment.

This protocol uses the MIDI message format, which is not used by the whole protocol. Most of the command mappings here will not be directly usable in the context of MIDI controllers. It can be parsed by standard Musical Instrument Digital Interface (MIDI) interpreters. Besides, some of the message data are used in a different way. Table 4.1 shows the different commands and their interpretations.

To perform the communication through Firmata protocol, the first task is to upload a standard Firmata sketch in Arduino board. This helps the Firmata protocol to identify the data packet coming from the serial port. Now we can write the Firmata code on our personal computer so that the communication is successful. In following example, we provide a standard Firmata demonstration with processing language (see Chapter 8 for more detail about processing).

TABLE 4.1

Firmata2.6.0 Commands and Interpretation

Type	Command	MIDI Ch	First Byte	Second Byte
Analog I/O	0XE0	PIN#	LSB(bit 0–6)	MSB(bit7–13)
Digital I/O	0X90	PORT	LSB(bit 0–6)	MSB(bit7–13)
Report analog port	0XC0	Pin#	Enable/disable 1/0	N/A
Report digital port	0XD0	Port	Enable/disable 1/0	N/A

The steps are as follows:

- First of all, we have to connect the Arduino in slave mode.
- Upload standard Firmata on Arduino.
- Import Firmata library in the processing environment.
- The code for standard Firmata is discussed later.

There are several functions that are associated with the standard Firmata code. Some of them are as follows:

Void readAndReportData(byte,int,byte) function allows I²C requests that does not require to register because some of the devices only require an interrupt pin to signify that the new data is available.

Void checkDigitalInput(void) function checks all active digital input pin and monitors their state change. It also adds any serial output sequence using Serial.print().

void setPinModeCallback(byte pin, int mode) function sets the pins to the correct states and sets some relevant bits in 2-bit arrays that track PWM and digital I/O status.

On the processing side, you have to do the following tasks:

- Download the processing library for Arduino available at arduino. cc/playground/uploads/Interfacing/processing-Arduino-0017.zip.
- Now unzip and copy the folder Arduino in the library folder of processing sketch.
- Run the following code in processing.

```
import processing.serial.*;
import cc.arduino.*;
Arduino arduino;
int ledPin = 13;

void setup()
{
//println(Arduino.list());
arduino = new Arduino(this, Arduino.list()[0], 57600);
arduino.pinMode(ledPin, Arduino.OUTPUT);
}

void draw()
{
arduino.digitalWrite(ledPin, Arduino.HIGH);
delay(500);
arduino.digitalWrite(ledPin, Arduino.LOW);
delay(1000);
}
```

Here, `cc.arduino` is the standard Arduino package for Firmata. In the `setup()` function, we have to create the object of Arduino class and pass the 0 element of the list vector that signifies the default port where Arduino is connected via PC. The baud rate should be passed as 57,600.

Now to initialize pin, we can call `pinMode()` functions of Arduino.

In the `draw()` function, to blink an LED using Firmata, we use `Arduino. digitalWrite()` method with pin and LOW/HIGH parameters.

4.5 Concepts Covered in This Chapter

- Basic functions and their use in Arduino
- Different classes of user-defined and library functions
- Firmata library and its use
- Accessing Arduino using Firmata

References

1. Coplien, Jim. "Advanced C++ programming styles and idioms." In *Tools*, p. 352. IEEE, 1997.
2. sparkfun. Retrieved from https://learn.sparkfun.com/.
3. Evans, Martin, Joshua J. Noble, and Jordan Hochenbaum. *Arduino in Action*. Manning, New York, 2013.
4. Arduino. *Explore Our Questions*. Retrieved from https://arduino.stackexchange. com/.

5

Conditional Statements

5.1 If-Else Conditional Statements

If-else is a kind of conditional statement that is primarily used to check whether a condition is true or false [1]. If the condition corresponding to the if() is satisfied, then the code corresponding to if block will execute. Generally, if block looks like the following:

```
if( condition){
                // code segment
                }
```

The condition within if block may be of different types. It might be some logical operation made within if statement.

The logical operators such as == (equals), >=, <=, and != are generally used for logical comparison. Sometimes, some logical && and || operators are also used to perform a comparison between two or more values.

In a general case, an if statement generally takes the value as either true or false. In compiler, it evaluates like 0 for any false condition and any non-zero value for true condition. Therefore, the following statement evaluates as true:

```
int x=5,y=6;
if(x&&y)    {  // evaluates as 1
Serial.println("hello");
}
```

Once the condition is satisfied, hello will be printed on the serial monitor. On the other hand, the following condition returns a false value:

```
int x=5,y=0;
if(x&&y)    {   // evaluates as 0
Serial.println ("hello");
}
```

5.1.1 Else

In most of the cases, if conditional statement is mainly associated with if-else conditional statement [2]. That means, if in any case if condition will not satisfy, the control directly goes to the hand of else part. The example is shown as follows:

```
int x=5,y=0;
if(x>6&&y>7)    {     // as the condition is not satisfied so it
will jumps the control to else part
Serial.println ("hello");
}
else{
Serial.println ("bye");
}
```

5.1.2 If-Else-If Ladder

If-else-if ladder can also be applicable for Arduino programming environment. In this case, if multiple if-else conditions are associated with the statement, then they can be branched through such kind of ladder-like mechanism.

The following example shows a demonstration of how we could form an if-else-if ladder format:

```
if(condition){
// code
}
else if (condition)
{
 // code
}
else{
  // code for else
}
```

5.2 Switch-Case Statement

Arduino supports switch-case statements. It is an elementary menu-based coding technique. In Arduino, the use of switch case is seldom. However, in some specific cases, switch-case statements may be helpful for choosing a certain task to be completed in the Arduino system by performing switch-case operation.

The following example shows the standard syntax for switch case:

```
switch(value){
 case con1:{
```

```
                         // statement 1
                             Break;
                         }
case con2 :{
                         // statement 2
                         Break;
                         }
default : {
                         // statement 3
                   Break;
                     }
}
```

In Arduino, the standard rule for switch case is similar to C or C++ language. For example, if we omit the break statement for cases, all the cases including default will execute one by one, but as the break is incorporated, the statement corresponding to case completes its execution, and it will immediately come out from the switch-case block.

Another major thing to remember while using switch case is that case constant has never been a floating point type. If we use the floating point as a case constant, it will always give us an error message.

5.3 Loops in Arduino

While, do-while, and for loops are supportable in Arduino. Mostly, the loop, in this case, performs repetitive tasks. In various scenarios such as controlling a servomotor or grabbing sensor data within a certain range of value, we generally use loops.

5.3.1 For loop

This is the most flexible loop that is often used to design the program in both Arduino and other languages such as C, C++, and Java. The statement of the for loop is shown as follows:

```
for (data type initialization; condition; counter)
```

In the first part, we have to initialize the loop counter with certain data type and value. In the middle part, it will check the boundary condition. In the third part, there will be the counter that will either incrementally or detrimentally update. The example is given as follows:

```
for( int i=0;i<10;i++){
    Serial.println(i,DEC);
  }
```

The above-mentioned program performs serial printing from 0 to 9 values in this case. For loop is a highly flexible loop in comparison with any other loops such as while or do-while because of easy understandability of the loop.

There are various representations made for for loop. Sometimes, the initialization statement can be made outside the loop. Moreover, the loop counter can be accommodated within the body of the loop sometimes. A for loop sometimes may have more than one boundary condition checking. We can accommodate them by adding a comma between two boundary conditions. A for loop can be made an infinite loop by simply removing the boundary condition. Sometimes, for loop can be left entirely blank like this for (;;). Doing so, for loop becomes infinite, and the iteration will go on and on. Such kind of loop may lock the program execution and go to the busy waiting state.

Sometimes loops are intentionally terminated with some; (semicolon) statement. Doing so, the loop will not do anything. In such case, the loop will execute within it, but the statement corresponding to the loop will never get executed. The following example illustrates the case:

```
for(int i=0;i<10;i++);
Serial.println("HELLO");    // hello will be executed only once
in this case.
```

5.4 While and Do-While Loops

5.4.1 While Loop

It is an entry-controlled loop in which the loop conditions will be checked during the entry of the loop. The syntax is shown as follows:

```
while (condition){
                The task to be performed by while

            }
```

While loop generally has a condition in it. The counter initialization is performed outside the loop, and the counter increment or decrement is generally performed in the body of the loop. The example is shown as follows:

```
int   x=10;
while( x<100){
Serial.println(x,DEC);
x++;
}
```

5.4.2 Do-While Loop

It is a special kind of loop often known as an exit control loop. Because the condition is checked during the time of exit of the loop, the name is so. The main property of do-while loop is that the loop will execute at least once even if the condition is not satisfied at a particular moment. The do-while loop basically ends with a semicolon, which suggests that the control is at the end of the loop, so no further statement is inside the loop in this case. An example of a do-while loop is shown as follows:

```
int x=1;
do{
    x++;
    Serial.println("hello");
  } while (x<=10);
```

In general practice, for loop and while loop are popularly used in program design. Do-while loop is seldom used by the program. Only in such a condition where the statement should execute at least once, the do-while loop is appropriate.

5.5 loop() in Arduino

loop() function fundamentally performs the looping operation in the Arduino environment. This is the most essential component of Arduino through which sensor data can be grabbed in a repeated fashion [3]. Generally, loop() function iterates at a constant time interval, but we can control the time duration of the execution of the loop function by applying a delay() function within it.

Sometimes interrupts() function can be accommodated in the loop to restart the interrupt that has been disabled by the noInterrupts() function. Interrupts are generally used to allow certain tasks to happen in the back end that is enabled by default. Some of the important features might be disabled as an interrupt is disabled [4]. Interrupt might disrupt the timing of the execution slightly.

```
void setup() {}

void loop()
{
  noInterrupts();
  // code that are time sensitive and critical should be
placed here
  interrupts();
  // other necessary codes
}
```

attachInterrupt() is a method that attaches the interrupts with the digital pin through which a device gets hosted. The pin number, interrupt service routine, and interrupt mode are the parameters of the attachInterrupt() function.

5.6 Concepts Covered in This Chapter

- Control statement fundamentals
- Looping
- loop() function and interrupt
- Switch-case operation

References

1. Monk, Simon. *30 Arduino Projects for the Evil Genius*. McGraw-Hill Education, New York, 2013.
2. Monk, Simon. *Programming Arduino Next Steps: Going Further with Sketches*. McGraw-Hill, New York, 2014.
3. Electronics Hub. *A Simple DIY Universal Remote Using Arduino*. Retrieved from https://www.electronicshub.org/.
4. The Engineering Projects. Retrieved from https://www.theengineeringprojects.com/.

6

Arduino Input Systems

Numerous input systems supported by the Arduino microcontroller board. The fundamental input types for Arduino are analog and digital inputs. The sensors that are mostly connected with Arduino are perhaps either digital types of sensors, such as ultrasonic sensor, tilt sensor, gyroscope, global positioning systems (GPS), and magnetometer, or analog sensors, such as temperature and humidity sensor, and infrared (IR) sensor. Interfacing the sensors as input is pretty challenging sometimes because of the proper calibration factor and proper input formatting techniques. Figure 6.1 shows the different input channels of Arduino.

6.1 pinMode() Function

pinMode() function is the basic function that is mainly used to specify the pin to behave as an input or output [1]. In general, Arduino pins are input

FIGURE 6.1
Input channels of Arduino.

by default, so we do not need to give an explicit declaration of the pin mode as input. By default, the configuration of the pin mode generally is in a high impedance state. The input signal is basically equivalent to $100\,M\Omega$ in series resistance in front of the pin itself. It means that it will pull very little current to drive the input from one state to another and can make the pins useful for the tasks such as driving a capacitive touch sensor. Reading an LED as a photodiode, or reading an analog sensor with an RCTime.

6.2 INPUT and INPUT_PULLUP Configuration

Often, it is necessary to steer up the input pin into a known state if no input is present. This can be performed by add implementing a pull-up register (to +5 V) or pull-down register to the input (a register to ground). Generally, $10\,k\Omega$ resistances are pretty good as a pull-up/pull-down resistance.

A $20\,k\Omega$ pull-up resistance is inbuilt in the ATmega microcontroller chip. It can be accessed through software tools. This built-inpull-up resistance can be accessed by setting up the pinMode() as INPUT_PULLUP. It efficiently inverts the behavior of the INPUT, which signifies HIGH when the sensor is in the off mode and LOW when it is in the ON mode.

The resister value mainly depends upon the microcontroller version that is used. In most of the AVR-based microcontrollers, this value is in between 20 and 50 $k\Omega$. When we connect as a sensor with a pin having INPUT_PULLUP, another end must be connected to the ground. Pull-up resistance passes enough current to drive an LED or an analog sensor.

The pull-up resistors are controlled by the same register that controls whether a pin is HIGH or LOW, and it is situated in the chip memory location. Consequently, a pin that is configured to have pull-up resistors turned on only when the pin is used as an INPUT. This will have the pin configured as HIGH if the pin is then switched to an OUTPUT mode with the pinMode() function. This works in the other direction as well, and an output pin that is left in a HIGH state will have the pull-up resistors set if switched to an input with pinMode().

Pins configured in OUTPUT with pinMode() are generally in a low-impedance state. This means that it can provide a significant amount of current to other circuits. ATmega pins can be thought as a source (provide positive current) or sink (provide negative current) having the range up to 40 mA (milliampere) of current to other circuits, device, and sensors. This produces enough current to brightly light up an LED (don't forget the series resistor), or run many sensors, for example, but not enough current to run most relays, solenoids, or motors.

Short circuits on Arduino pins, or attempting to run high current devices from them, can damage or destroy the output transistors in the pin or damage the entire ATmega chip. Often, this will result in a "dead" pin in the microcontroller but the remaining chip will still function adequately. For this reason, it is a good idea to connect OUTPUT pins to other devices with 470 Ω or 1 kΩ resistors, unless the maximum current drawn from the pins is required for a particular application.

6.3 `digitalRead()` Function

The main task of `digitalRead()` function is to read data in two states. It may be either HIGH or LOW (in terms of 1 and 0). We put the pin number as a parameter of the digital read function. It mainly reads the sensor that sends digital PWM signal to the microcontroller. It can also be invoked to read the state of a toggle switch or push button. Mostly, the `digitalRead()` function is associated with `loop()` function, because most of the time this function has to be called from the `loop()` function so that it can repeatedly read the sensor data.

```
int pBtn = 10;

// the setup function executes at once

 void setup() {

// start serial communication at 9600 baud rate.

 Serial.begin(9600);

  // assign push button as an input device

  pinMode(pBtn, INPUT);
}

// the loop function will run infinitely

void loop() {

 // read input pin

 int b_state = digitalRead(pBtn);

 // print out the state of the push button

 Serial.println(b_state);

 delay(1); // delay in millisecond to reads for stability

}
```

In the above-mentioned sketch, we are interfacing a push button that is connected in digital input line 10. `void setup()` routine, in this case, is used to declare the mode of the pin which is input in this case, and the serial communication also begins in this routine. Generally, we choose the 9,600 baud rate for serial communication. Next, in the `loop()` function, we are taking the input state by calling `digitalRead()` function. After getting the state of the button (either 0 or 1), we are now sending that data to the serial port using the `Serial.println()` function.

6.4 `analogRead()` Function

The primary function for reading analog signal from outside world is the `analogRead()` function. The analog signal is nothing but a continuous time-domain signal. There are various sources present to create an analog signal. Some popular analog devices are a potentiometer, analog sensors such as a simple photoresistor, an LM35 temperature sensor, a soil moisture probe, and many more.

In Arduino UNO, six analog pins are available: A0–A5. In case of Arduino Mega, a total of 16 pins are available, namely, A0–A15. The `analogRead()` function reads the analog information and returns it as a numeric digital value. The type of the data is an integer type. The Arduino has 10 bits of A/D converter which will quantize the input voltage level to five distinct voltage states. The integer value produced by this is in between 0 and 1,023, which means a total of 1,024 values. Thus, the resolution of the A/D converter is 0.0049 V per unit. The resolution and input range can be changed using the `analogReference()` function. Figure 6.2 shows the input pattern for analog sensors.

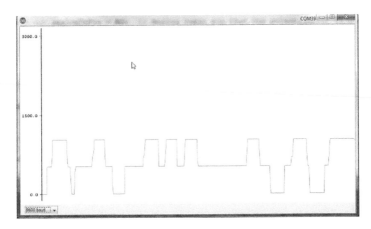

FIGURE 6.2
Analog input pattern.

6.5 Inputs for Firmata Library

Firmata generally talks to the computer system using USB communication channel. The application like processing communicates to it via the serial library. The serial communication is basically character oriented, and once it is properly setup, it is a good and reliable media of communication. Fermata is MIDI-based and is 3 bytes long. There are various input and output functions that are available for communication. To set up a digital output, we have to build the data from the parameter values of each bit. The reading of digital input involves three basic steps: The pin must be put into the digital mode, the port must be set to the input mode, and the decoding has to be done. As the port read mode is enabled, Arduino starts sending values to the appropriate pins. One of the properties of this protocol is that it will send or resend data when it changes. If the message is identical to the previous message, it will never be sent. Parsing method is used to do this task.

Some of the message sending and receiving are mentioned functions shown as follows:

`sendAnalog(byte pin, int value)` sends an analog message or analog signal to a specific pin. `sendDigitalPort(byte portNumber, intportData)` sends the single digital signal to a 8-bit port. The second method is `sendString(const char* string)`. This method specifically sends the strings to the host computer. Next, `sendString(byte command, byte bytec, byte *bytev)` is basically used to send a string to a host computer with a specific custom command. Another method named `sendSysex(byte command, byte bytec, byte* bytev)` sends a command with an arbitrary set of bytes. Finally, `write(byte c)` is generally used to write a byte data to the stream.

Several input functions that are considered to perform the input operation are as follows:

- available() is used to check whether any incoming message is received at the input buffer or not.
- The processInput() function, on the other hand, processes the incoming message taken from the buffer. It will also send the data to any registered callback method.
- Two other popular functions that are incorporated are attach(byte command, callbackFunction Function1), and detach(byte command), which are primarily used to attach a function with the incoming message type and detach a function from the incoming message.

In Firmata, the communication with Arduino can be done with the serial object. The serial baud rate, in this case, is 57,600 baud. Eight data bit frames are available with one stop bit, but no parity is required. The first task is to setup an Arduino to match the fundamental requirement. The settings define the mode and direction of I/O pins. Firmata defines various pin functions as pin mode. Some of the vital pin modes are 0 for digital in, 1 for digital out, 2 for analog input, 3 for PWM, and 4 for servo control. To set the pin mode, set a message [244 pins #mode]. It is easy enough to set a patch to map an arbitrary pin. As the port is set now after a little time, the setting has to be set.

Reading digital pin generally involves three steps. First of all, the pin has to be put into a digital input pin mode. The port must be enabled for input, and the data received must be decoded. As the port read is enabled, Arduino board is now ready to start sending values. As the value changes, the value will be immediately sent to the port. The values that are sent may be identical with the output data in output port, and here goes the challenge to separate the data with a process called parsing. Several parsing techniques are available. In one technique, the data come as a stream of numbers, which will be tested whether they are greater than some threshold value (127 in this case) and therefore a status will be sent.

Parsing of the analog message is quite similar with the digital parsing. The pins are generally set up to analog input mode in this case. The desired pins are enabled and the input data get parsed. Analog pins, in this case, have to be labeled as 0–5. Analog data has to be reported with a status message with a unique status number (typically 224). Two data bytes have to be combined to get a data value that ranges from 0 to 1,023.

6.6 Input Device Interfacing

Some pins of Arduino can be configured as both input and output [2]. In this section, we are going to mention the exact activity of the pins as they act as digital input or output pins. Some of the properties of the input and output pins are discussed next.

6.6.1 Input Pin Properties

As we know that the default nature of the Arduino pins is input, there is no need to declare it as input. The pins that are configured as input are generally in high impedance state [3]. It has a very low demand on the circuit and is actually equivalent to a 100 Ω series resistance in front of the input. These input pinstake very little current to move from one state to another. The operations include reading a photodiode, a capacitive touch sensor, and analog sensors like LM35 temperature module.

Although, logically, `pinMode(pin, INPUT)` produces a not connected state, sometimes an abrupt change of pin state is reported due to electrical noise picking from some adjacent circuitry or capacitive coupling with nearby pins.

Often, we do steer the input pin into a known state if the input pin is open. We can do this by using a pull-up or pull-down resistor. Pull-up resistors are connected with +5 V and pull-down resistors should ground to 0. The 10 K resistors are a good example of a pull-up or pull-down resistor.

The internal pull-up resistor value of the ATmega chips is 20 kΩ. It can only be accessed from the software. This can be done by performing `pinMode()` as INPUT_PULLUP [4]. This basically inverts the mode, that is, HIGH will produce an OFF state and LOW will produce an ON state.

The pull-up value changes as the microcontroller changes. Most of the AVR-based microcontrollers have a pull-up value between 20 and 50 kΩ. Some of them are in between 50 and 150 kΩ.

Pull-up registers generally draw enough current to drive an LED to glow perfectly.

Sometimes, attempting to run a device that takes high current damages and destroys the output transistor pin or ATmega chip. This causes a dead pin in ATmega chip although the entire microcontroller is in the functional mode. Sending a clear message reflects that the data chunk is clear. Since each message has a unique status number, it is pretty easy to identify a message that holds a specific route.

6.7 Concepts Covered in This Chapter

- Input methodology
- Analog input techniques
- Digital input techniques
- Input device interfacing
- Input pin properties

References

1. Barrett, Steven F. *Arduino Microcontroller Processing for Everyone!*. Morgan and Claypool Publishers, San Rafael, CA, 2010.
2. Kadam, Rahul, Ruchika Shelke, Pritesh Bagde, Gunwanta Bande, Shraddha Dongre, Hemnat Chachane, and B. E. Student. "Automated gate of animal cage using Arduino Mega 2560 and Raspberry Pi." *International Journal of Engineering Science* 17249 (2018): 16204–16206. http://ijesc.org/.

3. Caruso, Massimo, Antonino O. Di Tommaso, Rosalba Miceli, G. Ricco Galluzzo, Pietro Romano, Giuseppe Schettino, and Frank Viola. "Design and experimental characterization of a low-cost, real-time, wireless AC monitoring system based on ATmega 328P-PU microcontroller." In *AEIT International Annual Conference (AEIT), 2015*, pp. 1–6. IEEE, 2015.
4. Banzi, Massimo, and B. Massimo. *Getting Started with Arduino (Make: Projects)*. O'Reilly, Sebastopol, CA, 2008.

7

Arduino Output Systems

Various sets of output mechanism is supported by Arduino Microcontroller system. Various components get connected with Arduino as output devices sometimes directly or using some external shields. In this chapter, we are going to view and demonstrate the possible output mechanism for Arduino with some of the standard libraries as well as some of the well-known output devices and their interfacing mechanisms.

7.1 `digitalWrite()` Function

One of the most useful functions for the output of Arduino is the `digitalWrite()` [1]. It writes a HIGH or LOW value to the digital pin. The output nature of the pin will be decided by the `pinMode()` function. This makes the pin as an input pin or output pin depending on the ease of use. As the pin becomes an output pin, +5 V (+3.3 V for some specific 3.3 V board) will be the output for HIGH signal and 0 V for the LOW signal. In case of `pinMode()` as input, the `digitalWrite()` will activate or deactivate the input pull-up resistor. Mostly, it is recommended in case of pin mode as input, and the mode should be addressed as INPUT_PULLUP to activate the pull-up resistors. Sometimes, we call `digitalWrite()` as HIGH without making `pinMode()` as OUTPUT. Doing so, as we use any output device with it, like an LED, it will appear dim. This is because, as we haven't mentioned pin mode as OUTPUT, it enables the pull-up resistor and acts as a large current-limiting resistor. The following example shows the output setup for a `digitalWrite()` operation:

```
void setup(){

        pinMode(12,OUTPUT);
        }

void loop(){
            digitalWrite(12, HIGH);
            }
```

7.2 analogWrite() Function

7.2.1 Operation

analogWrite() is another function that performs the output operation of Arduino [2]. This function sends a PWM wave to the mentioned pin. This is mostly used to drive an LED light for fading purpose (where the brightness varies). To control the speed of the motor, the analogWrite() function is used. The fundamental principle of the function is that, for a particular duty cycle, the analogWrite() function will generate a steady sequence of the square wave. This continues until and unless a next invoke of analogWrite() or digitalWrite().

In case of Arduino UNO and similar boards, the frequency of the steady wave is 980 Hz for pins 5 and 6, and 490 Hz for other pins. Depending on the microcontroller, the analogWrite() function works in different pins. In case of ATmega168 and ATmega328P, pins 3, 5, 6, and 9–11 support this function. In case of ATmega2560 and Mega ADK44-46, pins 2–13 are dedicated for the same. Pins 9–11 are supported by ATmega8 only. Unlike the analogWrite() function, DACs are also used as true analog output. In Arduino DUE, not only pins 2–13 but also DAC0 and DAC1 support the analog output. One of the advantages of analogWrite() is that we may or may not use the pinMode() function to mention the OUTPUT for a given pin.

```
int d =0, apin = A0;
void setup(){

    pinMode(6,OUTPUT);

}

void loop(){
    d = analogRead( apin);
    analogWrite(6,d/4);
    }
```

Generally, the analog input pins A0–A5 (for UNO) take the value between 0 and 1,023, where the analogWrite() function performs an operation of 0–255 states. Sometimes, pins 5 and 6 produce a PWM which is of a higher duty cycle than expected. This happens due to the interaction with millis() and delay() functions. As a result, the 0 value never fully turns off the device connected to it.

7.2.2 PWM details

Pulse width modulation is a method used to get an analog signal in the form of a digital pulse. In this case, digital control device produces the square wave that represents the ON and OFF switching states. This switch also physically makes a transition from 0 to 5 V. The duration of this transition time, in this case, is considered to be the pulse width. To obtain a varying pattern of analog signal, we could increase the speed of switching or decrease it. If the switching speed is extremely high, the result will be at a constant steady voltage level, representing the full glow of the LED. Figure 7.1 represents the variable width of the PWM duty cycle.

Figure 7.1 depicts the frequency of the `analogWrite()` pulse. The value of the pulse ranges from 0 to 255. `analogWrite(255)` represents 100% duty cycle, whereas `analogWrite(127)` represents a 50% duty cycle and so on.

In the standard procedure of call of `digitalWrite()` or `analogWrite()`, we have a full control of the duty cycle of the PWM pulse. However, one disadvantage is that any interrupt applied to the processor affects the time, and hence, the width gets disturbed. While the processor is busy in other works, the output pin cannot be left with other works. It is also utterly difficult to determine the constant for a particular duty cycle as well as the frequency. In this case the manual count of the cycle is the only way.

7.2.3 ATmega PWM Access

We can have more access to the PWM compared to the `analogWrite()` function by controlling the PWM output. Actually, ATmega168/328 has three inbuilt timers: Timer0, Timer1, and Timer2. Each of them has two compare

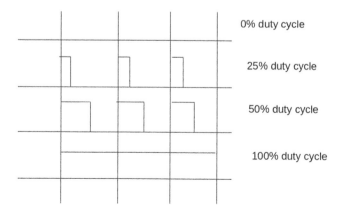

0% duty cycle

25% duty cycle

50% duty cycle

100% duty cycle

FIGURE 7.1
The duty cycle of the PWM signal for `analogWrite()`.

registers and two outputs. The compare register controls the width of the PWM signal. When the compare register reaches the timer value, the output will get toggled immediately. Two outputs of the timer basically have the same frequency, although the duty cycles of the timer are different.

Each of the timer comprises a prescaler that generates the timer clock pulse by dividing with the prescaler factor. The timer also generates an interrupt or match against the compare register. The timer comprises different PWM modes. The range of the timer values varies from 0 to 255.

- Fast PWM mode is the simplest mode where the timer value is rapidly counted from 0 to 255. The output turns on when the timer value is 0 and the timer is off when it matches with the compare register [3].
- Phase-correct PWM mode is a unique mode where the timer counts from 0 to 255, and after that, it will come down to 0. Output turns off as it hits and matches the compare register. The output of this is more symmetrical. In comparison with the fast PWM mode, the output frequency is just half.
- OCRA-based fast PWM is another class of PWM that is fundamentally based on the output control register. Timer2-based PWM operates on pins 3–11. A special mode called waveform generation mode is used here, which is set to 111 initially. In case of ORCA2, the top limit is assigned arbitrarily as 180.
- Fast PWM with varying time top limits is the method in which the timer counts from 0 to the ORCA, which is the value of the output compare register. After doing this, the timer will count from 0 to 255 and gives a significant control over the output frequency.

Figure 7.2 shows ATmega328 PWM pin configuration.

Consider a scenario to blink an LED by applying ON and OFF to it. If it is the case, the ATmega will create a square wave, and for a certain period of time, the signal is HIGH; after that, it should become LOW. In such case, if the on value gets divided by the off value, then multiplied by 100%, we obtain the actual result for the duty cycle (DC_t). Therefore, it can be written as

$$DC_t = \left\{ On_t / (On_t + Off_t) \right\} \times 100 \qquad (1)$$

Here, On_t is the ON time and Off_t is the OFF time duration.

Therefore, if the controller is on for 1 ms and off for 2 ms, then we can achieve a duty cycle for 33.33%. The output voltage, therefore, is computed by the following equation:

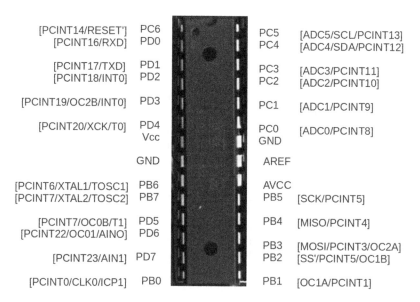

FIGURE 7.2
ATmega328 PWM configuration.

$$V_o = DC_t \times V_i \qquad (2)$$

Here, V_o is the output voltage and V_i is the input voltage.

From the above-mentioned equation, it is clear that if we have +5 V input and the PWM duty cycle is 25%, then the output analog signal voltage is 1.25 V. Since the prescaler acts as a counter, it allows to run the device in varying frequencies. In this way, it helps those devices that are sensitive to the varying PWM speeds. For example, a brushless motor generates heat at a very high PWM rate, and the motor will jitter if the PWM is too slow. As we discuss the different PWM modes, let us discuss the different mathematical models for them.

The fast PWM mode is a standard mode where the PWM acts as a normal counter. As the controller receives the signal, it sets the register TCINTn. If a match detects, an OCFnx flag get set and the signal is then sent to waveform generator. However, the frequency F_t of the PWM wave, therefore, can be computed as

$$F_t = \left[\text{Clock} / \{PS_v \times (1 - T_v)\} \right] \qquad (3)$$

Here, PS_v is the prescaler value, T_v is the top value, and Clock is the clock speed.

7.3 An Input–Output Example

A typical embedded device can take the input from the sensors and generate the output in various ways. One of the most popular amongst them is the visualization of output data. Several tools are available for visualization. One of the platforms is processing, where we can do the visualization by means of several abstract graphical designs that are also highly used for ambient and translucent lighting using LED and LCD screens. Another way of visualization is the plotting of the dataset generated by the sensor through the bar or line plotting. In this section, we represent the sensor data visualization using a plotting technique. To do that, we may use Python programming and the packages that are related with graphical visualization (more about Python has been discussed in Chapters 10–21). Another aspect of the output we have to realize in this case is the logging of the sensor data. We can do the same by using the Python language and its associated packages. In this section, we demonstrate the fundamentals of the output data logging and visualization. In this case, we have taken an ultrasonic echo sensor and take the distance data as input. The output data will log into an Excel sheet. We do use python openpyxl package for logging the data and pyserial package for serial port data receive purpose. The data can be accessed using a package called pandas. The task of pandas is to load data from excel and other multidimensional resources. After that, the loaded data will be visualized using the package called Mathplotlib. We have used a standard ultrasonic sensor code for Arduino to get the ultrasonic distance and print it in the serial port. The following code demonstrates the distance of data fetching using an ultrasonic sensor. The sensors trig pin should be connected with pin 9 of Arduino, and Echo pin should be connected with pin 10. Vcc and Ground should also be connected with +5 V and ground pin, respectively. The connected device is shown in Figure 7.3.

```
// code curtsy https://howtomechatronics.com/
const int tPin = 9;
const int ePin = 10;
    // defines variables
    long dur;
    int dist;
    void setup() {
    pinMode(tPin, OUTPUT);
    pinMode(ePin, INPUT);
    Serial.begin(9600);
    }
    void loop() {
    digitalWrite(tPin, LOW);
    delayMicroseconds(2);
    digitalWrite(tPin, HIGH);
```

FIGURE 7.3
Arduino connected with an echo ultrasonic sensor.

```
delay(10);
digitalWrite(tPin, LOW);
duration = pulseIn(ePin, HIGH);
dist= dur*0.034/2;
Serial.println(dist);
}
```

In the above-mentioned code, serial data will be supplied by the Serial.
print(dist) function. The second code segment has been designed
to take the serial data using the Python serial interface. The serial.
Serial(port,baud) function takes the serial data where the port name
and baud are supplied as parameters. As we get the data, we have to open
the workbook and invoke the active sheet. In this case, we typically take
300 data samples from the serial input and put them into Column 2 of Excel
worksheet. Column 3 will store the iteration number. Then, we have to save
the dataset. Now our task is to load the data and print it as a dataset representation. It can be done using pandas as pd object. The pd.read_excel()
function here reads the dataset of Excel format. Then, we just print the head
of the data by calling head(number of items). Now, we can plot the
data values of the head by invoking the plot() function. The m.head(297).
plot() function here does that task. Here "297" is the number of tuple present in the head. plt.show() finally shows the graphical representation of
the data. Figure 7.4 shows the dataset created by the head() function, and
Figure 7.5 shows the plot of the sensor data.

```
Data Written successfully !!
head is
       Unnamed: 0   4225\n    0
0             NaN     199     1
1             NaN     194     2
2             NaN     192     3
3             NaN     225     4
4             NaN     193     5
5             NaN     197     6
6             NaN     226     7
7             NaN     225     8
8             NaN     194     9
9             NaN     226    10
10            NaN     225    11
11            NaN     226    12
12            NaN     191    13
13            NaN     225    14
14            NaN     226    15
15            NaN     226    16
16            NaN     226    17
17            NaN     198    18
18            NaN     230    19
19            NaN     227    20
>>>
```

FIGURE 7.4
The dataset generated by pandas.

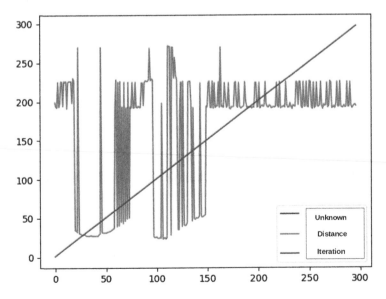

FIGURE 7.5
Data visualization.

```
from openpyxl import Workbook  # to create Excel spreadsheet
import serial          # to invoke serial interface
import pandas as pd # Excel data access
import matplotlib.pyplot as plt # plotting dataset.

s = serial.Serial('COM13',9600) # COM Port for serial data input
wb = Workbook() # creation of excel work book
ws = wb.active        # Active sheet

start_row = 2
start_column = 2
addl_column = 3
i = 0
while (i<300)  :
    ws.cell(row=start_row, column=start_column).value =
str(s.readline())
    ws.cell(row=start_row, column=addl_column).value = i
    start_row += 1
    i+=1

wb.save("dataset.xlsx")
print "Data Written successfully !!"

f = "dataset.xlsx"
m = pd.read_excel(f)

print 'head is'
print (m.head(20))

m.head(297).plot()
plt.show()
```

Figure 7.4 shows the dataset generated by the ultrasonic distance sensor. We have stored the number of iterations and the raw sensor distance data grabbed by the ultrasonic sensor. Figure 7.5 illustrates the visualization of the ultrasonic sensor data, where the *x*-axis represents the number of iterations of the void loop() function and the *y*-axis represents the distance. The first graph showing the iteration has a linear value, and the second graph showing the distance varies according to the motion of the object. The third value is unknown, arising from the blank column of the Excel sheet.

7.4 Concepts Covered in This Chapter

- Function of digitalWrite()
- Function of analogWrite()

- Arduino PWM characteristics
- Python support for output visualization

References

1. Margolis, Michael. *Arduino Cookbook: Recipes to Begin, Expand, and Enhance Your Projects*. O'Reilly Media, Inc., Sebastopol, CA, 2011.
2. Javed, Adeel. *Building Arduino projects for the Internet of Things: Experiments with Real-World Applications*. Apress, Berkeley, CA, 2016.
3. Monk, Simon. *Programming the Raspberry Pi: Getting Started with Python*. Mcgraw-Hill, New York, 2013.

8

Arduino with Processing

Arduino is a stand-alone open source microcontroller unit that serves various purpose [1]. Often, an Arduino board has been used for acquisition of sensor data or as a node that can host several sensing application. The data gathered from various sensor resources mostly kept in a permanent memory. This can be done using a standard Arduino shield, or sometimes, it can be stored in a local computer system or a cloud storage. The sensor data may be further visualized by means of various visualization tools. The term processing is quite relevant to Arduino, as it is popularly used for sensor data visualization accompanied with Arduino. Processing is a sketchbook based programming platform for visualization and to produce visual arts. It originated in 2001 first, and since then, it has been promoting software literacy within visual arts and literacy within real-time and embedded technology. There are various phenomena in which processing is so popular.

- It supports interactive programming paradigms with 2D, 3D, as well as portable document format (PDF) output.
- It supports a portable platform for Linux, Windows, MacOS X, and even android as well as the ARM.
- It has huge library support.
- It is compatible with multiple languages.
- Overall, it is an open source platform that can be freely downloadable.

8.1 Overview of Processing Language

8.1.1 General Overview

Processing is a language of visualization and visual arts [2]. The base language of processing is Java. Although Java has some of the syntactical features, it also consists of the phenomena of PostScript and OpenGL for visualization and the realization of the graphic elements; the OpenGL is basically incorporated for 3D visualization. A large group of languages are fundamentally associated with processing such as Arduino, wiring framework, C++, Java, and Android. Besides, processing support for JavaScript and HTML5 is also incorporated by means of the processing.jsframework [3,4].

Lots of people use processing every day. Day by day, the community support for processing increases and becomes stronger. The processing can be useful for various classes of people. People like visual artists, graphic designers, animators, and those who want to visualize sensor data in graphical form in real time or may want to design visual effects using sensor data can use processing.

8.1.2 Processing Software

As the software itself is open source, it is available in processing.org/. The stable versions for Windows, Linux, and MAC OS X are also available. Besides, the IDE is also available for Android in Play Store. In the Android version, one can create wallpapers and different virtual reality apps. We can easily run the codes in any tablet or smartphone without changing or partly changing the code itself. We also need not worry about installing the Software Development Kit or editing the layout. Processing for Android also lets the Android Application Programming Interface (API) to read sensor data or even exporting sketch as a signed package that is ready to upload in Google Play Store. For more detail about processing and Android, join Github (https://github.com/processing/processing-android).

Other supported platforms for processing are Windows 64 bit, Linux 64 bit, and Linux ARMv6hf. The older release can also be built form the source as well. The processing window is depicted in Figure 8.1.

In recent days, Python mode processing is also available. Processing that was released initially preferably used Java-based syntax, and the graphical phenomenon takes the inspiration from OpenGL and PostScript. The alternative language support for processing emerged by getting the language support of Ruby, Python, and JavaScript (shown in Figure 8.2).

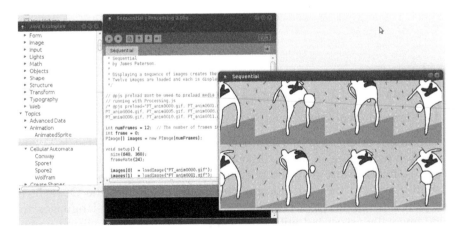

FIGURE 8.1
Processing window displays the basic animation.

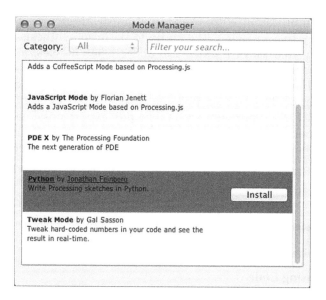

FIGURE 8.2
Different modes in processing.

To start processing development, first install processing IDE from .zip or .exe source and open the processing window. As we open the window, in the top right corner you can see the mode option. By default, the mode is Java mode. We can further change the mode from Java to Pythonmode.

As the Python mode is installed, you can check the basic sketch by typing the following:

```
def setup():
    size(600, 600)
    colorMode(HSB)
    noStroke()


def draw():
    fill(0x11000000)
    rect(0, 0, width, height)
    fill(frameCount % 255, 255, 255)
    ellipse(mouseX, mouseY, 20, 20)
```

It creates some basic shapes of rectangle and ellipse form with the color supplied.

Python mode is basically engineered over Java and designed to be compatible with the existing ecosystem and processing library. Some of the important libraries used in processing are Serial library that performs a serial communication with an external hardware device (USB, RS232), Sound library that processes the audio to perform synthesis and effects, Video library that plays a video file and creates movies, and Network library that sends and receives data in a network in the client–server mode. DFX export library creates DFX files to save geometry to load into third-party application. It comprises polygon, boxes, spheres, and other triangle base objects. PDF Export library makes pdf export possible. Hardware I/O library performs interaction with hardware such as Raspberry Pi or other Linux-based system. More about libraries are discussed further in Section 8.3.

8.2 Code Structure

8.2.1 Processing Code

As we install the processing environment on our computer, our task is to make some programs. In windows environment, we can open the IDE by double-clicking the icon. If the package is a .zip extension, we simply extract it at any directory and run the executable file. In case of Linux, to execute it, we must write ./processing (as it is a shell script having a .sh extension). Before that, we must change the mode of the processing executable file to the executable mode. We must give the command chmod 777 processing to do so. Once it is successful, it will open the IDE.

The fundamental code structure of processing is basically similar to the Arduino language, as it consists of void setup() function, and instead of having a void loop, there is a void draw() function. The job of void draw() function is to draw the pixel within the given image area. In our first program, we will simply draw some circles within a fixed frame area. The code is shown as follows:

```
void setup() {
size(500, 520);
smooth();
}
void draw() {
if (mousePressed) {
fill(1);
} else {
fill(252);
}
ellipse(mouseX, mouseY, 70, 70);
}
```

The output of the code will show the group of the circle in the given frame area of 500, 520, and as we move the mouse pointer over the area, more circles will be generated.

One of the major advantages of the processing application is that you can directly export your code into an executable application software. In such case, you have to click the export button. It will immediately create the application as per your requirement. By default, it will create the Windows 32/64 bit, Linux 32/64 bit, and MAC OS X application.

8.2.2 Drawing the Objects

Processing consists of two main functions to run the application. The first one is `void setup()` that is mainly used for initializing the variable and objects. It also initializes the size of the display frame of the window. The next method is `draw()`. This method runs like an applet life cycle, which means the code within this method actually executes in a cyclic fashion, until and unless the display window stops by clicking the close button.

To start drawing, the basic coordinate system of a computer screen and display frame need to be understood. Generally, the coordinate system comprises the X, Y axis grid of pixels. Any shape or size can be drawn using the coordinates mentioned. The function `size()` is often used to set the display area width. By default, the size is 100×100 pixels if we do not invoke the size function explicitly. The size function is actually used for setting the resolution of the visualization in the screen area.

A very fundamental drawing function is `point()`function. The parameter of this function is the x, y-axis. This is a basic function to realize the visualization of the pixels in the screen. We can use `point()` function to draw straight horizontal or vertical lines by simply changing the value of x- and y-axes within a loop.

Processing includes a large number of basic shape functions. Some of the functions are listed as follows:

- line(x1,y1,x2,y2)
- triangle(x1,y1,x2,y2,x3,y3)
- quad(x1,y1,x2,y2,x3,y3,x4,y4)
- rect(x,y,w,h)
- circle(x,y,w,h)
- arc(x,y,w,h,start,stop)
- ellipse(x1,y1,x2,y2)

The example demonstration for the earlier functions is illustrated in Figure 8.3.

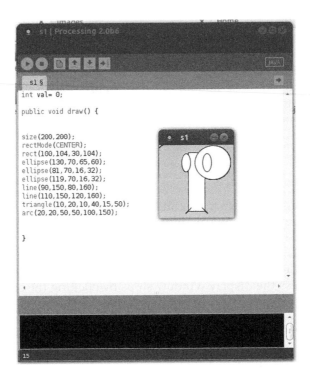

FIGURE 8.3
Implementation of basic shapes.

8.3 Libraries and Functions

Various libraries are incorporated in the processing environment. The most effective serial library is discussed next.

8.3.1 Serial Library

Serial library is one of the popular libraries that can send data from processing through a serial port like RS232 [5]. Most of the microcontroller boards such as Arduino and PIC are able to communicate with processing by using this library. Various functions are associated with this library. Some of them are discussed as follows:

- Serial: It is a class in Serial Library that actually controls all the necessary features of serial communication. There are a good number of functions associated with this Serial class that is shown later.
- available(): This function returns a number of bytes in integer and shows how many bytes are available at a particular instant of time.

- `read()`: This function is used to read the data byte from the read buffer. Reading should be done in the form of 0–255 values. If no data is available, then it will return–1. There are various other forms of the `read()` function available, such as `readChar()`, `readByte()`, `readByteUntil()`, `readString()`, and `readStringUntil()`. Until function generally reads data byte until a specific number of characters. If that character is not present, then null is returned. We have to put the destination character in this case as a parameter of the function.

- `buffer()`: This function is mainly used for creating a temporary buffer memory for storage. We have to pass the size of the buffer in this case. The syntax looks like `Serial.buffer(size)`. A `bufferUntil()` function is also associated with `buffer()` function that ensures the buffer position until a serial event is called.

- `Last()`: This method returns the last byte in the buffer available. If it is not available, then it immediately returns–1.

- Some input–output functions are also available. Those functions are `write()`, `clear()`, `stop()`, and so on. `write()` function is used to write on the serial port, `clear()` function clears out the port itself, and `stop()` function stops further communication on this port. There is another function known as `list()`.

- Hardware I/O: It is mostly used to access Raspberry Pi or another Linux-based computer system. This is basically a peripheral interfacing library. Its functionality is similar to the Arduino platform.

A very useful class in this library is General Purpose Input/Output GPIO class. This library comprises various features as follows:

- Network: It sends data through the network using the simple client–server method.

- Sound: It performs playback of the audio file and produce synthesized sound and effects.

- Video: It reads the movie from the camera and play videos.

- DFX Export: It is able to create DFX files and save geometry, which can be used by other applications

- PDF Export: It creates and exports PDF files with high-resolution vector graphics as well.

8.4 Visualization of an LDR Output

An LDR is a light detection sensor. It is basically a light-sensitive resistor [6]. The resistance values get changed as the light intensity changes. We can visualize the light intensity by capturing the light intensity data from the serial port of the device. The data can then be taken by the processing platform to generate a graphical visualization. In this case, we can use an Arduino DUE and interface a voltage divider circuit along with a light sensor. The Arduino code for getting LDR data is shown as follows:

```
//Arduino Code

int apin =A0;
void setup() {
  // put your setup code here, to run once:

 pinMode(apin,INPUT);
 Serial.begin(9600);
}

void loop() {
  // put your main code here, to run repeatedly:

  int d = analogRead(apin);

  Serial.println(d,DEC);

}
```

The following code represents the processing example to visualize the movement of the circle, which depends upon the change of light intensity. Figure 8.4 shows the visualization of the light intensity.

```
// Processing code

import processing.serial.*;
Serial port;
float br=0;

void setup(){

  size(800,600);
  port =  new Serial(this,"COM6",9600);
  port.bufferUntil('\n');
}

void draw(){

  background(br-88,br-44,br);

  fill(255,255,0);
```

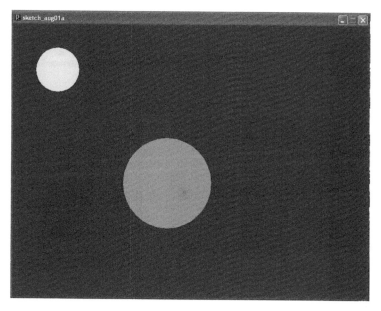

FIGURE 8.4
Visualization of light intensity.

```
  ellipse(100,100,br,br);
  fill(020,200, 77);
  ellipse(br+250,br+250,200,200);
  println(br);
}
void serialEvent(Serial port){
  br = float(port.readStringUntil('\n'));
}
```

8.5 Interaction with Arduino and Firmata

There is a strong compatibility between Arduino and processing. Often, we directly interact with processing and Arduino [7]. Sometimes, the interaction is to be controlled by the Firmata control protocol.

A fundamental example of Arduino processing communication is the sensor data visualization. The following example is a demonstration of how Arduino and processing work together.

In this example, we are trying to visualize the data of the ultrasonic sensor. The data has been grabbed by the ultrasonic sensor, and the distance has been computed using Arduino; further, the distance value will be sent to the

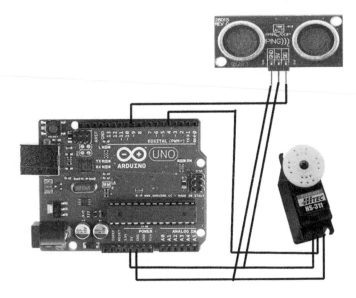

FIGURE 8.5
Connection diagram of Arduino-based radar system.

processing application, and it will visualize the data as a Radar like range finder. The system diagram is shown in Figure 8.5.

The following code demonstrates the Arduino and processing implementation, respectively:

Arduino Code

```
// Includes the Servo library
#include <Servo.h>.
// Defines Tirg and Echo pins of the Ultrasonic Sensor
const int trigPin = 3;//orange
const int echoPin = 2;//red white - vcc grey -gnd
// Variables for the duration and the distance
long duration;
int distance;
Servo myServo; // Creates a servo object for controlling the
servo motor
void setup() {
  pinMode(trigPin, OUTPUT); // Sets the trigPin as an Output
  pinMode(echoPin, INPUT); // Sets the echoPin as an Input
  pinMode(4,OUTPUT);//yellow
  pinMode(11,OUTPUT);
  Serial.begin(9600);
  myServo.attach(9); // Defines on which pin is the servo
motor attached on 9 purple
}
```

```
void loop() {
  // rotates the servo motor from 15 to 165 degrees
  for(int i=0;i<=180;i++){
  myServo.write(i);
  delay(30);
  distance = calculateDistance();// Calls a function for
calculating the distance measured by the Ultrasonic sensor for
each degree

  Serial.print(i); // Sends the current degree into the Serial
Port
  Serial.print(","); // Sends addition character right next to
the previous value needed later in the Processing IDE for
indexing
  Serial.print(distance); // Sends the distance value into the
Serial Port
  Serial.print("."); // Sends addition character right next to
the previous value needed later in the Processing IDE for
indexing
  }
  // Repeats the previous lines from 165 to 15 degrees
  for(int i=180;i>0;i--){
  myServo.write(i);
  delay(30);
  distance = calculateDistance();
  Serial.print(i);
  Serial.print(",");
  Serial.print(distance);
  Serial.print(".");
  }
}
// Function for calculating the distance measured by the
Ultrasonic sensor
int calculateDistance(){

  digitalWrite(trigPin, LOW);
  delayMicroseconds(2);
  // Sets the trigPin on HIGH state for 10 micro seconds
  digitalWrite(trigPin, HIGH);
  delayMicroseconds(10);
  digitalWrite(trigPin, LOW);
  duration = pulseIn(echoPin, HIGH); // Reads the echoPin,
returns the sound wave travel time in microseconds
  distance= duration*0.034/2;
  if (distance < 40)
    {
      digitalWrite(4,HIGH);
        digitalWrite(11,HIGH);

    }
```

```
   else
   {
     digitalWrite(4,LOW);
       digitalWrite(11,LOW);
   }
     return distance;


}
```

The Arduino collects the ultrasonic sensor data that is mounted on a servo-motor. The statement #include<Servo.h> actually attaches the functionality to control the servo in this case. Then, we have to declare trigger and echo pin as constant variables.

Now in the setup() function of Arduino, declare trigger pin as output and echo as input. The servo will get attached top in no. 9.

In the loop() function, a for loop is made first to rotate the servo from 15° to 165°. Here, myServo.write(i) function performs the movement of the servo, having a delay of 30 ms. As it moves, the calculateDistance() function is called to grab the data from the ultrasonic sensor and send it to the serial port, including the angle of the servo. Now the second for loop is design for the same purpose but for reverse repetition.

The job of the calculateDistance() function is to read the echopin and return the travel time of the sound wave in a microsecond. The distance value, in this case, is equal to duration X 0.034/2. If the value is less than 40, then we trigger the alarm light connected in pins 4 and 11.

The processing code for the same is shown next. We will explain the functionality of the code at the end of the program.

```
import processing.serial.*; // imports library for serial
communication
import java.awt.event.KeyEvent; // imports library for reading
the data from the serial port
import java.io.IOException;
Serial myPort; // defines Object Serial
// defubes variables
String angle="";
String distance="";
String data="";
String noObject;
float pixsDistance;
int iAngle, iDistance;
int index1=0;
int index2=0;
PFont orcFont;
void setup() {
```

```
  size (1200, 700);
  smooth();
  myPort = new Serial(this,"COM3", 9600); // starts the serial
communication
  myPort.bufferUntil('.'); // reads the data from the serial
port up to the character '.'.
                                         // So actually it
reads this: angle,distance.
}
void draw() {

   fill(98,245,31);
   // simulating motion blur and slow fade of the moving line
   noStroke();
   fill(0,4);
   rect(0, 0, width, height-height*0.065);

   fill(98,245,31); // green color
   // calls the functions for drawing the radar
   drawRadar();
   drawLine();
   drawObject();
   drawText();
}
void serialEvent (Serial myPort) { // starts reading data from
the Serial Port
   // reads the data from the Serial Port up to the character
'.' and puts it into the String variable "data".
   data = myPort.readStringUntil('.');
   data = data.substring(0,data.length()-1);

   index1 = data.indexOf(","); // find the character ',' and
puts it into the variable "index1"
   angle= data.substring(0, index1); // read the data from
position "0" to position of the variable index1 or thats the
value of the angle the Arduino Board sent into the Serial Port
   distance= data.substring(index1+1, data.length()); // read
the data from position "index1" to the end of the data pr
thats the value of the distance

   // converts the String variables into Integer
   iAngle = int(angle);
   iDistance = int(distance);
}
void drawRadar() {
   pushMatrix();
   translate(width/2,height-height*0.074); // moves the
starting coordinates to new location
   noFill();
   strokeWeight(2);
```

```
  stroke(98,245,31);
  // draws the arc lines

  arc(0,0,(width-width*0.0625),(width-width*0.0625),PI,TWO_PI);
  arc(0,0,(width-width*0.27),(width-width*0.27),PI,TWO_PI);
  arc(0,0,(width-width*0.479),(width-width*0.479),PI,TWO_PI);
  arc(0,0,(width-width*0.687),(width-width*0.687),PI,TWO_PI);
  // draws the angle lines
  line(-width/2,0,width/2,0);
  line(0,0,(-width/2)*cos(radians(30)),(-width/2)*sin(radi
ans(30)));
  line(0,0,(-width/2)*cos(radians(60)),(-width/2)*sin(radi
ans(60)));
  line(0,0,(-width/2)*cos(radians(90)),(-width/2)*sin(radi
ans(90)));
  line(0,0,(-width/2)*cos(radians(120)),(-width/2)*sin(radi
ans(120)));
  line(0,0,(-width/2)*cos(radians(150)),(-width/2)*sin(radi
ans(150)));
  line((-width/2)*cos(radians(30)),0,width/2,0);
  popMatrix();
}
void drawObject() {
  pushMatrix();
  translate(width/2,height-height*0.074); // moves the
starting coordinates to new location
  strokeWeight(9);
  stroke(255,10,10); // red color
  pixsDistance = iDistance*((height-height*0.1666)*0.025);
// covers the distance from the sensor from cm to pixels
  // limiting the range to 40 cms
  if(iDistance<40){
    // draws the object according to the angle and the
distance
  line(pixsDistance*cos(radians(iAngle)),-pixsDistance*sin(rad
ians(iAngle)),(width-width*0.505)*cos(radians(iAngle)),-
(width-width*0.505)*sin(radians(iAngle)));
  }
  popMatrix();
}
void drawLine() {
  pushMatrix();
  strokeWeight(9);
  stroke(30,250,60);
  translate(width/2,height-height*0.074); // moves the
starting coordinates to new location
  line(0,0,(height-height*0.12)*cos(radians(iAngle)),-(height-
height*0.12)*sin(radians(iAngle))); // draws the line
according to the angle
  popMatrix();
```

```
}
void drawText() { // draws the texts on the screen

  pushMatrix();
  if(iDistance>40) {
  noObject = "Out of Range";
  }
  else {
  noObject = "In Range";
  }
  fill(0,0,0);
  noStroke();
  rect(0, height-height*0.0648, width, height);
  fill(98,245,31);
  textSize(25);

  text("10cm",width-width*0.3854,height-height*0.0833);
  text("20cm",width-width*0.281,height-height*0.0833);
  text("30cm",width-width*0.177,height-height*0.0833);
  text("40cm",width-width*0.0729,height-height*0.0833);
  textSize(40);
  text("Akshay6766 ", width-width*0.875,
height-height*0.0277);
  text("Angle: " + iAngle +" °", width-width*0.48,
height-height*0.0277);
  text("Distance: ", width-width*0.26, height-height*0.0277);
  if(iDistance<40) {
  text("          " + iDistance +" cm", width-width*0.225,
height-height*0.0277);
  }
  textSize(25);
  fill(98,245,60);
  translate((width-width*0.4994)+width/2*cos(radians(30)),
(height-height*0.0907)-width/2*sin(radians(30)));
  rotate(-radians(-60));
  text("30°",0,0);
  resetMatrix();
  translate((width-width*0.503)+width/2*cos(radians(60)),
(height-height*0.0888)-width/2*sin(radians(60)));
  rotate(-radians(-30));
  text("60°",0,0);
  resetMatrix();
  translate((width-width*0.507)+width/2*cos(radians(90)),
(height-height*0.0833)-width/2*sin(radians(90)));
  rotate(radians(0));
  text("90°",0,0);
  resetMatrix();
  translate(width-width*0.513+width/2*cos(radians(120)),
(height-height*0.07129)-width/2*sin(radians(120)));
  rotate(radians(-30));
```

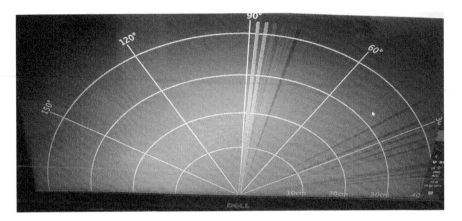

FIGURE 8.6
The processing radar screen.

```
text("120°",0,0);
resetMatrix();
translate((width-width*0.5104)+width/2*cos(radians(150)),
(height-height*0.0574)-width/2*sin(radians(150))));
  rotate(radians(-60));
  text("150°",0,0);
  popMatrix();
}
```

The above-mentioned code creates the visualization that is similar to radar scanning (Figure 8.6). To get that visualization, the size of the screen is mentioned as 1,200×700. Then, the serial port is opened as a port object by invoking myPort = new Serial(this,"COM3", 9600). The port can also be invoked by another call Serial.list()[0]. In this case, Serial.list()[] returns the port name. By default, it is COM1. It can be changed by incrementing the index of Serial.list()[0]. Then, myPort.bufferUntil('.') is invoked, which reads the data from the serial port up to the character ".". So, it actually reads angle and distance.

8.6 Mouse Handling in Processing

In processing, we can handle the mouse activity as shown in Figure 8.7 [8]. It is a simplified event handling mechanism in processing. There are numerous functions that are involved with mouse programming that can

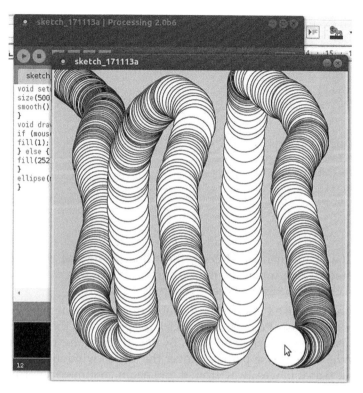

FIGURE 8.7
Mouse activity control.

control the mouse activity efficiently. As processing is based on Java, fundamental activities of the mouse are solely controlled by the Java event handling tools and interface at the back end, and since the processing produces a level of abstraction, the functions (methods) that are defined by processing generally invoke Java at the back end. We are going to discuss some of the major mouse handling methods of the processing in the following section.

1. mouseClicked(): This function is mainly invoked after a mouse button has been pressed and then released. Generally, mouse or keyboard event is associated with the draw() function. If we do not use it under draw(), it is simply invoked once and stops listening to events. Two overloaded function syntax is generally in use: The first one is mouseClicked() with no parameter and the other one is mouseClicked(event) with an event object generated by the program.

An example is shown as follows:

```
int val= 0;

public void draw() {
    fill(val);
    rect(20, 20, 60, 60);
}

void mouseClicked() {
    if (val == 0) {
        val = 255;
    } else {
        val = 0;
    }
}
```

In the above-mentioned code, the value of the `val` variable gets shifted to zero under the mouse click event. In this case, the value will change to 255 if it is zero; otherwise, it is zero.

2. mouseX: mouseX is a system variable that consists of the X coordinate of a mouse. Processing can track the position of the mouse if the mouse is in the current window. By default, the value of the same is 0. When the mouse pointer is over the sketch window, it holds the value 0 as it moves away from it and again continues reporting the updated position.

The following code shows the utility for the same:

```
public void draw() {
    background(225);
    line(mouseX, 24, mouseX, 70);
}
```

3. MouseY: This system variable stores the vertical position of mouse movement. Other properties of this variable are similar to the mouseX system variable.

4. pmouseX: It is a system variable that holds the previous frame position of the current frame position. We can find different values of pmouseX and pmouseY when it is referenced inside the events such as mousePressed(), mouseMoved(), and draw(). Inside draw(), both pmouseX and pmouseY variables are updated only once per frame. In the mouse event, they are generally updated each time the event is invoked. Some latency may be observed if the values are not updated immediately, resulting in chopping interaction and delay. If those variables get updated, multiple time per frame results a lot of gap because of the changes of the value lot many times between the call to the line.

The following simple program prints the positional difference of mouse:

```
void draw() {
  background(205);
  line(mouseX, 30, pmouseX, 70);
  println(mouseX + " : " + pmouseX);
}
```

5. pmouseY: It is a system variable that comprises Y-direction or vertical direction of the previous frame of the mouse current location. All other functionalities are similar to the pmouseX system variable.

6. mousePressed(): This function is somehow similar to mouse-Click() and responds only once after each time the mouse is pressed. A mouse button variable associated with this is generally used to track which button has been pressed.

7. mouseButton: It is a system variable that stores the mouse button that has been pressed. Generally, it grabs LEFT, RIGHT, and CENTER values depending upon the left, right and center buttons pressed in the mouse. If no button event is found, it will show a 0 value for the same. For this reason, the mousePressed() function should be called first to check the status of pressing and then check the value of the system variable. The following code shows an example of that:

```
void draw() {
  rect(20, 20, 40, 40); // drawing of the rectangle
}

void mousePressed() {
  if (mouseButton == LEFT) {      // check and performs actions
    fill(3);
  } else if (mouseButton == RIGHT) {
    fill(252);
  } else if (mouseButton == CENTER) {
    fill(116);
    }
  else{
    fill(178);
    }
}
```

8. mouseReleased(): This function calls every time the mouse button gets released. The event generated by the mouseReleased() function can, therefore, be used to perform a different operation such as drawing or filling of colors. Operationally, it is actually similar to mouseClicked() and mousePressed() functions.

9. `mouseMoved()` and `mouseDragged()`: The `mouseMoved()` function is invoked as the mouse button is not pressed but the mouse is moved physically. The `mouseDragged()` function, on the other hand, responds as the mouse is dragged. We can do any specific operation under these functions.

```
int mv = 0;

void draw() {
  fill(mv);
  rect(20, 20, 45, 45);
}

void mouseMoved() {   // changes the mv value across the image
  mv += 4;
  if (mv > 255) {
    mv = 0;
  }
}
```

10. `mouseWheel()`: This function is a mouse wheel control function and returns a positive value when the mouse wheel is down and returns a negative value as it goes far away. In this function, we have to pass the event as a parameter. We generally use the `MouseEvent` event object that controls the event functionality.

8.7 Colors in Processing

The color scheme of the processing comprises an 8-bit color for each color value; therefore, a 24-bit color scheme is maintained (as RGB). We generally use different functions to implement the color for some background as well as foreground designs. Some of the color functions that are most popularly used are shown as in Figure 8.8.

`background(color code)` function is a function that sets the background color of the display window. We can send any value between 0 and 255 as a parameter of the background function. The `stroke(color code)` function, on the other hand, sets the outline of the color stroke that is provided as a fill color. After setting the stroke value, we can setup `fill(color code)` function that fills the particular shape. We can accommodate a number of parameters in `fill()` functions like `fill(R, G, B)`. In this case, we set Red, Green, and Blue values explicitly. In the `fill()` function, we can set up the color code within 0–255 as well. A shape function is called thereafter

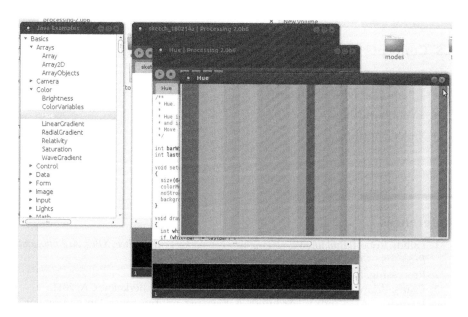

FIGURE 8.8
Color handling in processing.

to create a shape to visualize the shape with a predefined color. The activity of the `stroke()` and `fill()` functions can be eliminated by using `stroke()` and `fill()` functions, respectively.

Processing also has a color selector through which we can select the color visually. To use these, we can go to the Menu bar and click Color selector.

Color transparency is another parameter that we can set during fill. It is nothing but the percentage appearance of color for a given shape. It is often called an Alpha value. We should change the Alpha value for a particular color to make it see through. `Alpha` value should be the fourth value in the parameter, and the invocation looks like `fill(R, G, B, Alpha)`. The range of the Alpha value is also between 0 and 255, where 0 is for complete transparent and 255 is 100% opaque.

There are two standard modes of color scheme mainly used for color handling in processing. The most popular is an RGB mode that supports 24-bit color (32 bits including Alpha value), and we can use a function called `color mode()` in which we can put the mode including the values. The format is given as follows:

`color mode(RGB,100,200,30,120)` where the red value goes to 100, Green value goes to 200, Blue value goes to 30, and Alpha value goes to 120. Instead of using the RGB scheme, we can use the HSB technique to refer Hue, Saturation, and Brightness for the graphics.

8.8 Concepts Covered in This Chapter

- Drawing fundamental using processing
- Interaction with hardware
- Event handling
- Color handling

References

1. Faludi, Robert. *Building Wireless Sensor Networks: With ZigBee, XBee, Arduino, and Processing.* O'Reilly Media, Inc., Sebastopol, CA, 2010.
2. Processing. Retrieved from https://processing.org/.
3. Evans, Brian. *Beginning Arduino Programming.* Apress, Berkeley, CA, 2011.
4. D'Ausilio, Alessandro. "Arduino: A low-cost multipurpose lab equipment." *Behavior Research Methods* 44, no. 2 (2012): 305–313.
5. Kushner, David. "The making of Arduino." *IEEE Spectrum* 26 (2011): 1–7.
6. Reas, Casey, and Benjamin Fry. "Processing.org: A networked context for learning computer programming." In *ACM SIGGRAPH 2005 Web Program,* p. 14. ACM, 2005.
7. Reas, Casey, and Benjamin Fry. "Processing.org: Programming for artists and designers." In *ACM SIGGRAPH 2004 Web Graphics,* p. 3. ACM, 2004.
8. Reas, Casey, Ben Fry, and John Maeda. *Processing: A Programming Handbook for Visual Designers,* The MIT Press, Cambridge, MA, 2007.

9

Real-Life Code Examples

In this chapter, the application of Arduino and the related sensors has been emphasized. We primarily describes small piece of real-life projects and demonstrate the uniqueness of different applications. At the beginning, we had implemented a rail gate control in an automatic fashion. Later, a heartbeat monitoring, and an Liquefied petroleum gas (LPG) leakage detector will be addressed with an example coding.

9.1 Automated Rail Gate Control

We can design a simple rail gate control by using Arduino and some proximity sensor [1,2]. The system is designed in such a way that, as the train comes within a certain distance from the level crossing, the alarm buzzer will start ringing and the gate will be closed. As the train passes and travels to a certain distance, the zone becomes clear, and therefore, the gate is open and the buzzer stops. Pieces of equipment necessary to make this project are listed as follows:

1. Arduino UNO, mega
2. An IR proximity sensor
3. A 12- V DC geared motor
4. A buzzer
5. An L293D motor driver Integrated Circuits (IC) or Arduino motor shield
6. A base cardboard and a wooden rod

9.1.1 Stepwise Procedure

To make a prototype project, first of all, a track should be made over which we have to apply the sensor at a certain distance [3–5]. This can be done by applying hot glue or screw. The motor should be attached with rubber or glue at one side of the rail track. Now attach a wooden or plastic rod with the axis of the motor so that it can be used as a gate. Next, fix the IR sensor on both sides of the track, which should be in equal distance from the motor, to

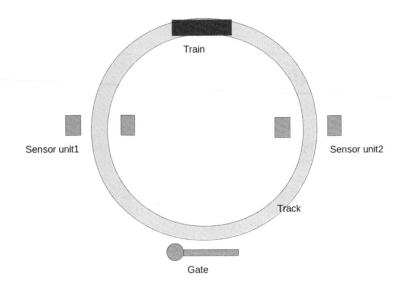

FIGURE 9.1
System diagram for rail gate controller.

show that one of them is used for sensing the arrival of the train, and accordingly, the gate should be closed. And the task of the second sensor is to open the gate as it passes through the second one. Figure 9.1 shows the fundamental diagram of the rail track. To make the sensor work properly, remove the LED of the sensor and just fix it opposite to the phototransistor attached in the sensor module.

9.1.2 Circuit Diagram

The circuit comprises the following components: (1) IR sensor that should be connected between Vcc and GND pin of Arduino, and (2) the output pin of Arduino that should be connected with pins 2 and 3, respectively. The motor driver (L293D) should be connected with Arduino pins 4 and 5. A motor connected with the motor driver is shown in Figure 9.2. Optionally, we can add a buzzer along with the motor to make a level crossing alert system.

The source code for the gate control is shown as follows:

```
int so_1=2;
int so_2=3;
int mo_1=4;
int mo_2=5;
void setup(){
pinMode(so_1,INPUT);
pinMode(so_2,INPUT);
pinMode(mo_1,OUTPUT);
```

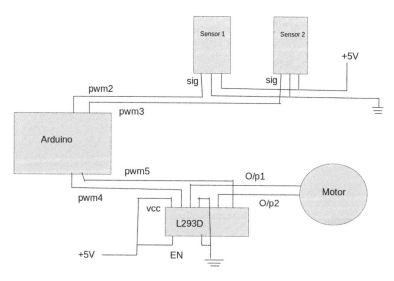

FIGURE 9.2
Circuit diagram.

```
pinMode(mo_2,OUTPUT);
}
void loop(){
Zp:
if(so_1==LOW){
digitalWrite(mo_1,HIGH);
digitalWrite(mo_2,LOW);
delay(600);
digitalWrite(mo_1,HIGH);
digitalWrite(mo_2,HIGH);
Xp:
if(so_2==LOW){
digitalWrite(mo_1,LOW);
digitalWrite(mo_2,HIGH);
delay(600);
digitalWrite(mo_1,HIGH);
digitalWrite(mo_2,HIGH);
delay(1100);
goto  Zp;
}goto Xp;
if(so_2==LOW){
digitalWrite(mo_1,HIGH);
digitalWrite(mo_2,LOW);
delay(600);
digitalWrite(mo_1,HIGH);
digitalWrite(mo_2,HIGH);
Yp:
if(so_1==LOW){
```

```
digitalWrite(mo_1,LOW);
digitalWrite(mo_2,HIGH);
delay(600);
digitalWrite(mo_1,HIGH);
digitalWrite(mo_2,HIGH);
delay(1100);
goto Zp;
}
goto Yp;
}
}
}
```

9.2 Arduino-Based Heartrate Monitoring System

The fundamental principle of heartbeat monitoring is the photoplethysmograph. According to this principle, the intensity change of the light passing through an organ is equivalent to the change of the volume of the blood in that organ [6].

The sensor for heartbeat monitoring is primarily an LED or sometimes an IR LED and a phototransistor or an LDR that detects the light/IR intensity. This kind of light source and sensor arrangement can be done in two different ways: a transmissive sensor and a reflective sensor.

In case of transmissive sensor, the light source and detector must be placed facing each other, and the finger of the person must be placed between the transmitter and the receiver.

Reflective sensor, on the other hand, has the light source and the detector adjacent to each other, and the finger of the person must be placed in front of the sensor. The heartbeat sensor consists of an LED or an IR LED and a photodiode. The signal is further received by photosensor and gets amplified with the amplification circuitry. Further, the signal is sent to the microcontroller for processing.

The main components used in this project are as follows:

- Arduino UNO
- 16 × 2 LCD display
- 10K pot
- 330 Ω resistor
- Push button
- Heartbeat sensor module
- Breadboard and connecting wire

The circuit design with Arduino is shown in Figure 9.3.

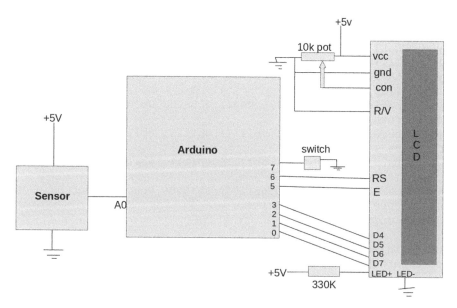

FIGURE 9.3
The circuit design for a heartbeat interface.

9.2.1 Working Methodology

The circuit diagram shows that the sensor has three pins: Vcc, GND, and signal pin. The signal will connect with analog pin A0. The raw data further get calibrated, and the final BPM value gets computed.

The value is then printed to a 16 × 2 LCD screen with certain values. As the system started, it is instructed to press the push button to start grabbing heartbeat. As it starts, it takes the pulse from finger to compute and display the heartbeat in beats per minute (BPM).

The output display will be interfaced with digital PWM pins 0, 1, 2, 3, 5, and 6. In pin 7, we will interface the push button. We have to interface the 10K pot with LCD screen to control the brightness of the screen.

The Arduino code is given as follows:

```
#include<LiquidCrystal.h>
LiquidCrystal lcd(6, 5, 3, 2, 1, 0);
int dat_1=A0;
int stat=7;
int co=0;
unsigned long t_mp=0;

byte cChar1[8] = {0b00000,0b00000,0b00011,0b00111,0b01111,0b01
111,0b01111,0b01111};
```

```
byte cChar2[8] = {0b00000,0b11000,0b11100,0b11110,0b11111,0b11
111,0b11111,0b11111};
byte cChar3[8] = {0b00000,0b00011,0b00111,0b01111,0b11111,0b11
111,0b11111,0b11111};
byte cChar4[8] = {0b00000,0b10000,0b11000,0b11100,0b11110,0b11
110,0b11110,0b11110};
byte cChar5[8] = {0b00111,0b00011,0b00001,0b00000,0b00000,0b00
000,0b00000,0b00000};
byte cChar6[8] = {0b11111,0b11111,0b11111,0b11111,0b01111,0b00
111,0b00011,0b00001};
byte cChar7[8] = {0b11111,0b11111,0b11111,0b11111,0b11110,0b11
100,0b11000,0b10000};
byte cChar8[8] = {0b11100,0b11000,0b10000,0b00000,0b00000,0b00
000,0b00000,0b00000};
void setup()
{
lcd.begin(16, 2);
lcd.createChar(1, cChar1);
lcd.createChar(2, cChar2);
lcd.createChar(3, cChar3);
lcd.createChar(4, cChar4);
lcd.createChar(5, cChar5);
lcd.createChar(6, cChar6);
lcd.createChar(7, cChar7);
lcd.createChar(8, cChar8);

pinMode(dat_1,INPUT);
pinMode(start,INPUT_PULLUP);
}

void loop()
{
lcd.setCursor(0, 0);
lcd.print("Place finger here");
lcd.setCursor(0, 1);
lcd.print("Now Start !!! ....");

while(digitalRead(stat)>0);

lcd.clear();
t_mp=millis();

while(millis()<(t_mp+10000))
{
if(analogRead(dat_1)<100)
{
co+=1;

lcd.setCursor(6, 0);
```

```
lcd.write(byte(1));
lcd.setCursor(7, 0);
lcd.write(byte(2));
lcd.setCursor(8, 0);
lcd.write(byte(3));
lcd.setCursor(9, 0);
lcd.write(byte(4));

lcd.setCursor(6, 1);
lcd.write(byte(5));
lcd.setCursor(7, 1);
lcd.write(byte(6));
lcd.setCursor(8, 1);
lcd.write(byte(7));
lcd.setCursor(9, 1);
lcd.write(byte(8));

while(analogRead(dat_1)<100);

lcd.clear();
}
}

lcd.clear();
lcd.setCursor(0, 0);
co*=6;
lcd.setCursor(2, 0);
lcd.write(byte(1));
lcd.setCursor(3, 0);
lcd.write(byte(2));
lcd.setCursor(4, 0);
lcd.write(byte(3));
lcd.setCursor(5, 0);
lcd.write(byte(4));

lcd.setCursor(2, 1);
lcd.write(byte(5));
lcd.setCursor(3, 1);
lcd.write(byte(6));
lcd.setCursor(4, 1);
lcd.write(byte(7));
lcd.setCursor(5, 1);
lcd.write(byte(8));
lcd.setCursor(7, 1);
lcd.print(count);
lcd.print(" BPM");
t_mp=0;
while(1);
}
```

9.3 Arduino-Based LPG Detector System

LPG is the most essential component used for domestic, industrial, and automobile purposes. The leakage of LPG, however, leads to a catastrophic disaster. Alert on LPG leakage, however, can minimize or sometimes reduce mishaps. In this project, we are going to develop an LPG leakage detection system that can be used in home or industries. The system is designed in such a way that as it detects any leakage, an alarm will be blown. The circuit of the system has the following components.

- Arduino
- LPG gas sensor module
- 1K resistor
- LCD display
- BC 547 transistor
- Battery pack
- Jumper wires
- Buzzer

9.3.1 Working Principle

An LPG gas detector module is used to detect LPG. When a leakage is detected by the sensor, it generates a HIGH pulse to its output pin. When Arduino receives a high pulse, it sends a message to the LCD, and the LCD shows a detection occurs message. The buzzer will immediately start the alarm. When the LPG sensor sends a LOW pulse to Arduino, it shows no LPG detected.

The modular components for the system are shown in Figure 9.4.

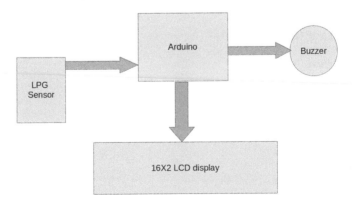

FIGURE 9.4
LPG-sensing application.

The sensor module is connected with the input line of the Analog input port of an Arduino sensor module consisting of MQ3 LPG sensor that detects LPG leakage with a high-precision MQ3 module that has an inbuilt heater inside. It requires some heat to sense the gas; therefore, it may require 10–15 min to go to the working condition. The buzzer and the LCD display module are connected to the digital pin of the Arduino. The sensor data coming from the gas sensor must be read by digitalRead function, and the LCD and buzzer have to be driven by some PWM signal using `digitalWrite` () function. The LCD screen module is attached to Arduino in a 4-bit mode. Read Write (RW), Register Select (RS) and Enable (EN) control pins are directly connected with Arduino pin2, and data pins are connected with pins 4, 5, 6, and 7, respectively.

```
#include <LiquidCrystal.h>
LiquidCrystal lcd(3, 2, 4, 5, 6, 7);

#define lpg_sens 18
#define buzz 13

void setup()
{
  pinMode(lpg_sens, INPUT);
  pinMode(buzz, OUTPUT);
  lcd.begin(16, 2);
  lcd.print("LPG Detection system !!!....");
  lcd.setCursor(0,1);
  lcd.print(" This is Test !!!....");
  delay(1900);
}

void loop()
{
  if(digitalRead(lpg_sens))
  {
    digitalWrite(buzz, HIGH);
    lcd.clear();
    lcd.print("LPG Leakage Found !!!......");
    lcd.setCursor(0, 1);
    lcd.print("   HIGH ALERT !!!!!  ..... XXX ");
    delay(300);
    digitalWrite(buzz, LOW);
    delay(400);
  }

  else
  {
    digitalWrite(buzz, LOW);
    lcd.clear();
    lcd.print("  No LPG  Detected.. ALL OK... ");
```

```
lcd.setCursor(0,1);
lcd.print("    Leakage Found .... HIGH ALERT.....!!!    ");
delay(900);
}
```

9.4 Conclusion

In this chapter, we have implemented a classical and popular example of automated gate control, heartrate monitoring, and gas-detecting application. The projects may further be modified as per the requirement of the users. The rail gate control project may further be enhanced with automated signaling system. We can make an automated lock gate control by influencing the rail gate control. The heartrate monitoring project has been further modified as the automated IoT-based electrocardiogram (ECG) application. These can be connected with remotely connected medical units. LPG-sensing application can be modified to a sensor grid-type application by considering the need of gas detection in industrial applications.

9.5 Concepts Covered in This Chapter

- Rail gate control application
- Heartrate monitoring application
- LPG detection application

References

1. Margolis, Michael. *Arduino Cookbook: Recipes to Begin, Expand, and Enhance Your Projects.* O'Reilly Media, Inc., Sebastopol, CA, 2011.
2. Mazidi, Muhammad Ali, Rolin D. McKinlay, Danny Causey, and P. I. C. Microcontroller. *Embedded Systems.* Pearson Prentice Hall, Upper Saddle River, NJ, 2008.
3. Meade, Travis. *Pulse Width Modulation Using an Arduino.* Michigan State University, East Lansing, MI, 2014.
4. Saha, Amit. "Learning to program the Arduino." *Linux Journal* 2011, no. 211 (2011): 2.
5. Oxer, Jonathan, and Hugh Blemings. *Practical Arduino: Cool Projects for Open Source Hardware.* Apress, Berkeley, CA, 2011.
6. Craft, Brock. *Arduino Projects for Dummies.* John Wiley & Sons, Inc., Hoboken, NJ, 2013.

10

Introduction to Python Language

10.1 Features of Python Language

Python is an open source language that is highly used in today's systems and software development. It is widely popular for intelligent computation. Although the fundamental structure of Python is somehow similar to FORTRAN, it is more powerful than that [1]. Most of the modern day computation including big data, machine learning, IoT, computer vision are the most common example of Python applications. Python is called an interpreted language rather than a compiled language. Python is also called dynamically typed language, which means if a variable is created, it won't be declared explicitly. There are many modern extraordinary features incorporated in Python language. Some of the most unique and appealing features of Python are as follows:

1. It is concise.
2. The block can be created without brace indentation.
3. Very rich collection types such as List, Tuple, Dictionary (often called Maps), and Sets are available.
4. List iterators are present like Java.
5. Simplified object-oriented support is also present.
6. Exception handling is present.
7. It is very powerful for scripting.
8. Several sequence types such as Strings and List are available.

One of the best features of Python is that it is a free and open source software. We can freely copy any module of the Python code and can modify the same. We can also study the source code of Python and modify according to our requirements.

10.2 Python Versions

Very common and popular versions of Python that are widely used nowadays are Python 2.7.x and 3.x versions. Some of the key features of 2.7.x version were changed in the Python 3.x version [2,3].

1. One of the major changes is the `print` statement [4]. In Python 2.7.x, we generally use `print` as a statement, whereas in Python 3.x, it is not a statement, but it simply becomes a `print()` function. Python 3 will encounter an error message if we don't write the statement inside parentheses. So the following statement works fine for Python 2.7 but encounters an error in Python 3:

   ```
   print 'Hello World!'
   ```

   so instead of that in python 3.x, we should write

   ```
   print('Hello World!')
   ```

   We can check and print the current version of Python by calling the `python_version()` function.

2. The second change that is quite dangerous sometimes if you are porting your code from the lower version is because this code never gives an error it computes the value wrongly.

   The following code in Python 2.7.x version gives different results compared to Python 3.x version:

   ```
   print '5 / 2 =', 5 / 2
   print '5 // 2 =', 5 // 2
   print '5 / 2.0 =', 5 / 2.0
   print '5 // 2.0 =', 5 // 2.0

   5 / 2 = 2
   5 // 2 = 2
   5 / 2.0 = 2.5
   5 // 2.0 = 2.0
   ```

   Python 3.x will produce an output like the following:

   ```
   print ('5 / 2 =', 5 / 2)
   ```

```
print ('5 // 2 =', 5 // 2)
print ('5 / 2.0 =', 5 / 2.0)
print ('5 // 2.0 =', 5 // 2.0)
5 / 2 = 2.5
5 // 2 = 2
5 / 2.0 = 2.5
5 // 2.0 = 2.0
```

3. In Python 2.x version, two types of strings are available: one type is ASCII `str()` and the other type is `Unicode()`. Byte type is not available in this version. However, in Python 3,UTF-8 type Unicode string is available. Besides, it also supports two byte types: byte and byte array.

```
print type("Hello this is test") statement produces
the following output in Python 2.7

<type 'str'>

print(' test", type(b' hello this is test")) statement
produces following output in Python 3.x

<class 'bytes'>
```

Python 3.x also supports Unicode in the following way:
```
print('Unicode utf-8 format \u03BCnico\u0394é!')

Unicode utf-8 format μnicoΔé!
```

4. `xrange()` function is widely used in Python 2.7.x for creating iterable objects. It generally doesn't run infinitely; therefore, it is exhaustive. In Python 3.x, the `xrange()` function gets abolished and the alternative `range()` function is introduced. In Python, `xrange()` produces a name error.

10.3 Python Installation

Since Python is an open source and free software, we can simply download it and run the executable installer [5,6]. Python can be downloaded free of cost, which is most advantageous. The versions of Python for different platforms are available freely in python.org.

10.3.1 Detailed Installation Procedure in Windows

In a Windows environment shown in Figure 10.1, we generally use Python installer for 64 bit (amd64) or 32 bit (X86) [7]. Versions such as 3.x are available in the form of .exe file. Some of the versions above 3.5 may be unstable while installing Service Pack 1 or less for Windows 7. The stable release of Python like Python 2.7.x is generally user-friendly and can be installed in any system like Windows XP, 7, or higher. The most recommended version is Python 2.7.14 that has good stability. These versions are available in.MSI format for Windows.

10.3.2 Installing Python in Ubuntu Linux

The simplest way to install Python in Ubuntu is using a command-line interface. In the interface (shown in Figure 10.2), we need to perform the following command:

```
sudo apt-get install python
```

10.3.3 Installation of Python Using the Source in Ubuntu

Before we install using .tgz file, we have to install some dependencies. To install dependencies, we should give the following commands:

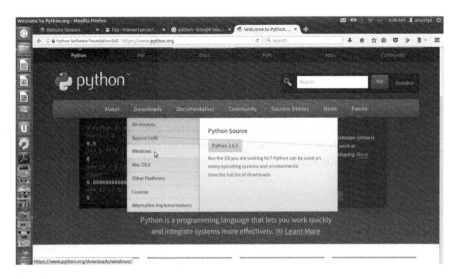

FIGURE 10.1
Python source.

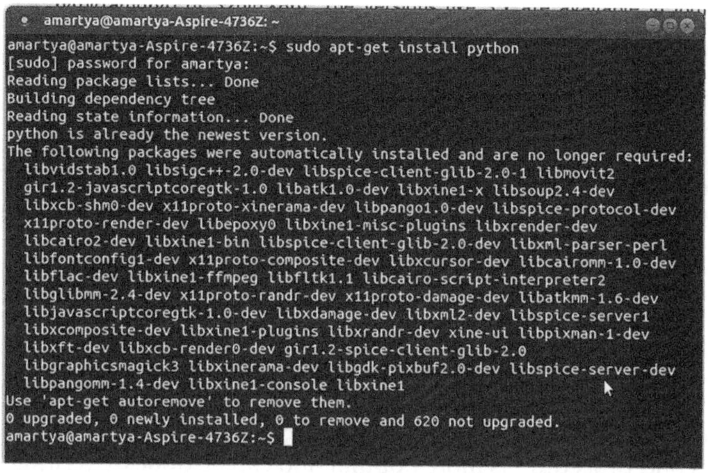

FIGURE 10.2
Installation in Ubuntu.

```
$ sudo apt-get install build-essential check install
$ sudo apt-get install libreadline-gplv2-dev libncursesw5-dev
libssl-dev libsqlite3-dev tk-dev libgdbm-dev libc6-dev
libbz2-dev
```

Then, the following command should be used for download:

```
version=2.7.13
cd ~/Downloads/
wget https://www.python.org/ftp/python/$version/Python-
$version.tgz
```

Now extract the file and go to the directory as follows:

```
tar -xvf Python-$version.tgz
cd Python-$version
```

We can install it using checkinstall or directly using command.

```
./configure
make
sudo checkinstall
```

10.4 Writing Some Basic Programs

There are two ways to run the Python code. In the first case, we can directly use command-line interface often called Python shell shown in Figure 10.3.

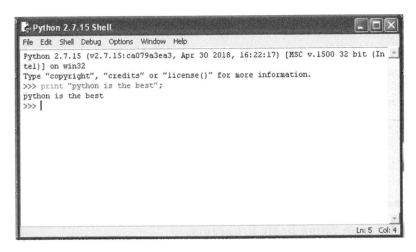

FIGURE 10.3
Execution in command-line interface.

This can be invoked from command-line interface terminal of Linux or Windows [8–10]. In Windows, IDLE interface is also available to run the code like command line interface.

The second procedure is more popular and can be done by writing Python scripts. Python scripts are the .py extension codes that can be interpreted by the Python interpreter. In Windows IDLE, there is a direct provision to run the code by pressing the F5/run button. In case of Ubuntu or Windows CLI, we simply use the command Python filename.py. Figure 10.4 presents the technique that is used to run in a Windows environment.

Figure 10.4 shows the direct execution of the Python code in the IDLE editor. In this case, we only type the code in the shell line by line, but this mechanism is unsuitable for writing big programs. Therefore, an alternate approach is the use of scripts that we generally save as filename.py. We can directly write through IDLE from File-> new script and then save it as Python file and simply run by pressing F5 as shown in Figure 10.5.

FIGURE 10.4
Direct execution of the program in shell.

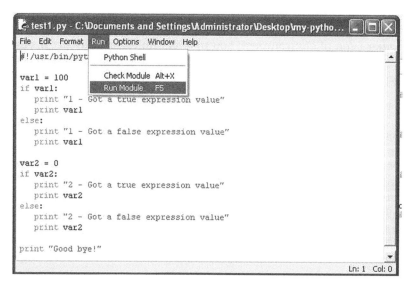

FIGURE 10.5
Execution of script in IDLE.

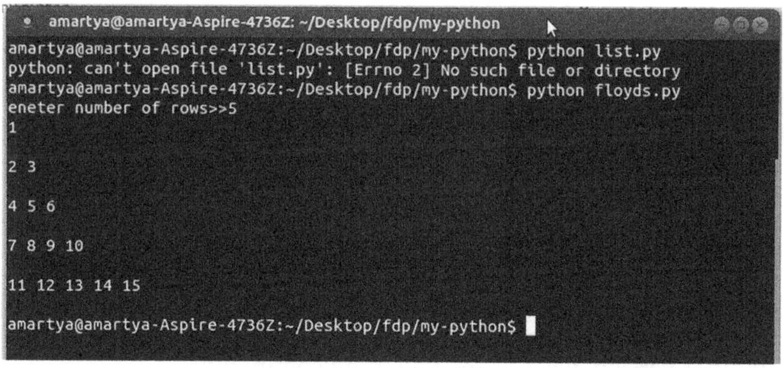

FIGURE 10.6
Execution of script in CLI (Ubuntu).

In case of Ubuntu Linux, the same procedure is applicable, but we have to write directly in the command-line interface as shown in Figure 10.6.

10.5 Installation Using pip

Before any further installation is done, you have to check further the version of the Python you are currently using. To see this in the command prompt, just type the following:

```
$python -version
```

In another type of environment, you may get an error message. If you are using Ipython or Jupyter, you can use the following command to check the version:

```
In [1]: import sys
        !{sys.executable} --version
Python 3.6.3
```

It is recommended to write {sys.executable} in place of Python to ensure that commands are running on the Python installation matching the currently running PC.

To check whether pip is available or not, you can use the following command:

```
pip -version
```

Python pip already gets installed in Windows in the Python/scripts directory. If it is in Linux, you may explicitly install pip. If it is not installed, you can do a bootstrap from the standard library by doing the following:

```
python -m ensurepip --default-pip
```

The most common use of pip is to install the package using some requirement specifiers. This comprises the name of the application with the version that is optional. Therefore, we give the following commands:

```
pip install 'application name'
to specify version we can give the command
pip install 'application name == 2.1'
```

To install a package that is greater than or equal to some version and less than another version, we give the following command:

```
pip install 'application name >= 2.1,<2.5'
```

To install a package compatible with the certain version, we can give the following:

```
pip install 'application name ~= 2.1.4'
```

Upgrading a package is also possible by passing the upgrade parameter. Also, we can install it from some specific user sites.

```
pip install - -upgrade application
pip install - - user application
```

We can even find the prereleased user packages and install them in our system by using the following command:

```
pip install - - pre-application package
```

10.6 Concepts Covered in This Chapter

- The basic idea about Python
- Installation of Python 2.7.14
- Installation of pip and another package
- Different versions and their comparison

References

1. Bernard, Joey. "Functions." In *Python Recipes Handbook*, pp. 49–54. Apress, Berkeley, CA, 2016.
2. Chun, Wesley J. *Core Python Applications Programming*. Pearson Education India, Tharamani, India, 2012.
3. Harms, Daryl D., and Kenneth McDonald. *The Quick Python Book*. Manning Publications, Greenwich, CT, 2000.
4. w3schools.com. Retrieved from https://www.w3schools.com/.
5. Liskov, Barbara, and Stephen Zilles. "Programming with abstract data types." In *ACM Sigplan Notices*, vol. 9, no. 4, pp. 50–59. ACM, 1974.
6. Lutz, Mark, and Mark Lutz. *Programming Python*. Vol. 8. O'Reilly Media, Inc., Sebastopol, CA, 1996.
7. Python. *Python 3.7.3 Documentation*. Retrieved from https://docs.python.org/.
8. Power, Russell, and Alex Rubinsteyn. "How fast can we make interpreted Python?"*arXiv preprint arXiv:1306.6047* (2013).
9. Rossum, Guido. *Python Reference Manual*. Amsterdam, The Netherlands, 1995.
10. Schildt, Herbert. *C/C++ Programmer's Reference*. McGraw-Hill, New York, 2000.

11

Operators, Variables, and Expressions

An operator plays a key role in developing the logics and expression in any programming language. All the operators perform some specific tasks to achieve the result. Some of the operators may perform more than one task too. In Python language, all the standard operators that are responsible for arithmetic, logical, and various relational operation can be directly applicable to operands. Operators such as arithmetic, unary, binary, logical, and relational are some common types of operators in Python. Python also has some special kind of operator syntax for some of the specific tasks that make the language distinguishable from other programming languages. In this chapter, we discuss the various operators that are used in Python language.

11.1 Operators in Python

In Python, various operators are used [1]. The operators are mainly applicable to data and make different expressions. In this chapter, we discuss different types of operators and different types of data used in the Python language.

The operators that are present in Python are similar to the C or C++ language. The operators are mainly categorized into the following types:

- Arithmetic operators
- Comparison operators
- Bitwise operators
- Logical operators
- Assignment operators
- Special operators
- Membership operators

11.1.1 Arithmetic Operators

The fundamental task of an arithmetic operator is to perform an arithmetic operation such as addition, subtraction, multiplication, or division [2,3]. Generally, the arithmetic operators are expressed as unary and binary operators. The binary operators are mainly acting upon two or more operands, and the unary operators are acting upon only one operand. Most popular types of the unary operators are unary – and unary +. The unary – operator is mainly used to represent a negative number.

Some additional arithmetic operators used in Python are "//"and "**". In case of the // operator, x//y represents the floor division of two numbers. As a result, it produces a whole number adjusted to the left in number line.

The following line describes the output of the // operator:

```
>>> x=10
>>> y=3
>>> print('x//y=',x//y)
('x//y=', 3)
>>>
```

x**y, on the other hand, produces xy, and the result is as follows:

```
>>> print('x**y=',x**y)
('x**y=', 1000)
>>>
```

11.1.2 Comparison Operators

Comparison operators are mainly used for logical comparison of different data members. While comparing, the operator returns the Boolean values as either true or false after evaluation. Two or more values can be compared by adding logical operators with a comparison operator. The following example returns the Boolean value for logical comparison:

```
x=10
y=3
print('x >= y is',x>=y)
```

The output, in this case, is "True."

Table 11.1 shows the different comparison operators.

11.1.3 Logical Operators

These operators perform a logical comparison between two operands. This operation can be done in unary and binary modes. The main logical operators used in Python are "and," "or," and "not." For any two given Boolean values, the and operator produces ANDing operation between two data. The corresponding statement is given as follows:

```
x=true
y=false
```

TABLE 11.1

Different Operators and Their Functionalities

Operator	Significance	Illustration
>=	Greater than or equal to, returns true if a>= b; otherwise, false	a>=b
<=	Less than or equal to, returns true if a<= b; otherwise, false	a<=b
==	Equal to, returns true if both the values are equal	a==b
!=	Not equal to, returns true if the values are not equal; otherwise, false	a!=b
>, <	Evaluates the value as true for greater than or less than, respectively	a>b a<b

```
print(x and y)
Results false
```

Logical operators are also applicable to numeric type values, but in that case, the evaluation depends on the occurrence of the data member. In the case of or operator, the result value is equal to the first member. The output shows the following:

```
>>> z= x or y
>>> print(z)
10
```

For and operation, the member should be the second operand.

```
>>> z=x and y
>>> print(z)
3
```

For unary operators such as "not," the evaluation performs based on whether the value is equal to zero or not. If the value is 0, it returns true; otherwise, false.

```
>>> print (not x)
False
>>> a=0
>>> print( not a)
True
```

11.1.4 Bitwise Operators

These operators are fundamentally acting upon bits, and they are unary as well as binary types of operators [4]. The operator first takes the binary form of any integer data and then applies the operation on it directly. The following statement is a demonstration of bitwise and that produces 0 as a result:

```
>>> x=2
>>> y=5
>>> print (x&y)
0
```

In case of x|y, it performs or operation of the data value bit by bit; therefore, the following statement produces 7 as output, because 5|7 implies (010)|(101) which is equal to 111 (7 in decimal).

```
>>> x=2
>>> y=5
>>> print (x|y)
7
```

Bitwise NOT operator is a unary operator that performs the complement of the value. Operationally, they are the same as C or C++ language. And it is symbolized as ~. Another important operator that is highly used in Python is bitwise XOR operator symbolized by ^. It performs exactly the same operation as XOR gate and is operationally similar to other languages.

Bitwise shift operators are another class of operators primarily used for left and right shift operations on bit values, respectively. The syntax is shown as follows:

X=x>>2 signifies that the bit pattern of x gets right shifted by two significant positions, and the final value, therefore, stores into x again. The same case can also be applicable to left shift operation as follows:

X=x<<2 signifies that the bit pattern of x gets left shifted by two significant digits. Generally, it is observed that shifting one significant digit to the left results in a data value is equal to multiplying the previous data by 2. For a right shift, this value gets divided by 2 for each significant digit's right shift.

11.1.5 Identity Operators in Python

Two special operators known as identity operators are available in Python. The syntax is "is" and "is not." Both of them will return Boolean values (True or False) based on the identity of the operators. The "is'" operator returns True if the two operands are identical. In case of any object, it returns true if the two operands represent the same object and False if they do not. The following code describes the operation of is and is not:

```
>>> x=7
>>> y=7
>>> print(x is y)
True
```

```
>>> print(x is not y)
False
>>>
```

If the values are a string, then also it will check the equality by observing the cases.

Therefore, if two strings are in different cases, then the is operator returns false.

For list type, this concept works based on the objects of the same reference. So, the following code segment will produce the output as false:

```
>>> x=[1,2,3]
>>> y=[1,2,3]
>>> print (x is y)
False
```

Since both x and y lists have the same values, but, as they belong to two different list objects, the result will be false in this case.

11.1.6 Membership Operators

The membership operators check whether a value is present in a particular string, tuple, list, set, or dictionary.

```
>>> x='test code'
>>> print('t' in x)
True
>>> print ('T' not in x)
True
```

The second illustration shows that, if we do apply 'in' operator for any dictionary, then the key of the value can match but not the value. Therefore, the following code performs the operation as follows:

```
>>> y= {1:'m',2:'n'}
>>> print(1 in y)
True
>>> print('m' in y)
False
```

11.1.7 Assignment Operators

Like other languages, = operator is used as an assignment operator in Python too. Operators such as +=, −=, /=, *=, **=, and //= are also supported. The assignment operator is used to assign not only a single variable value but also multiple variables.

```
a,b,c = 5,6,7
x,y,z = 9, 20.5,'abcd'
```

In the second case, the leftmost data goes to the leftmost variable and so on.

The swapping can also be performed by using the assignment operator easily. Consider two variables a = 5 and b = 6 in the normal programming language. Here we do swap them by using two different techniques.

```
Case 1:    t = a
           a = b
           b = t

Case 2:    a = a+b
           b = a - b
           a = a - b
```

But in the case of Python, we need not do it. Instead, we just simply write the following:

```
a,b = b,a
```

11.2 Clearing the Screen of the Console

Several methods are involved in clearing the IDLE or console screen in Python. One of the most popular and efficient brute force techniques is to write a definition for the clear screen by printing a certain number of newline characters using a print() statement. The code for the same is shown as follows:

```
>>> def cls():
        print('\n'*50)

>>> cls()
```

Here, we simply define the cls() function that consists of a print() statement. In the print() statement, we do print '\n' (newline character) 50 times. As a result, it produces 50 new lines and hence makes the screen clear.

11.3 Concepts Covered in This Chapter

- Arithmetic operators
- Logical operators

- Bitwise operators
- Identity operators
- Assignment operators
- Printing clear screen

References

1. Van Rossum, Guido, and Fred L. Drake. *PYTHON 2.6 Reference Manual*. CreateSpace, Scotts Valley, CA, 2009.
2. Van Rossum, Guido, and Fred L. Drake. *The Python Language Reference Manual*. Network Theory Ltd., Bristol, 2011.
3. Abrahams, David, and Ralf W. Grosse-Kunstleve. "Building hybrid systems with Boost. Python." *CC Plus Plus Users Journal* 21, no. 7 (2003): 29–36.
4. Lawson, Caleb. "CSC 415: Programming Languages Dr. Lyle November 4, 2014." http://campus.murraystate.edu/academic/faculty/wlyle/415/2014/Lawson.pdf (2014).

12

Decision-Making and Control Flow

Making decision in to a program code is the main aspect of the programming [1]. Decision-making and flow control are two primary components for developing software. In the decision control technique, one or more expressions should be evaluated individually or in parallel. Finally, the result returns either true or false; based on this, some pieces of statement or programming logic will be executed. Python interprets any nonzero or nonnull value as true and a zero or null value as false. The representation of the value of nonzero is presented as true and for nonzero, it is false.

12.1 `if` Statement

`if` statement is commonly used to evaluate the condition and the primary statement for condition checking [2]. Figure 12.1 shows the operational behavior of `if` statement.

The syntax of `if` statement is shown as follows:

```
a=20
b=30
if a<b : print(" a is smaller ")
```

`if` statement can hold multiple conditions to be checked at a time by using and and or operators. An `if` statement is always followed by a":".

```
if a<5 and a>1: print( "value is ")
                     print (a)
```

`if` statement is often followed by an `else:` statement so that the `if` statement gets branched conditionally. The following example shows the if-`else` conditional statement. Here, the `if` condition has been tested based on the value of var2. As the value of var2 is nonzero, it will generate a true value. So the `if` part will be executed in this case.

```
var2 = 10
if var2:
   print "1 - Got a true expression value"
```

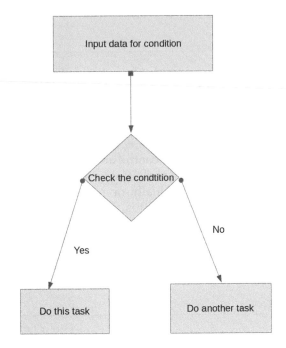

FIGURE 12.1
if conditional statement.

```
    print var2
else:
    print "1 - Got a false expression value"
    print var2
```

12.2 If-elif-else Ladder

12.2.1 Ladder Representation

In some cases, we do use if-elif-else statement to generate an if-elif-else ladder kind of formation. In such case, the if condition may contend some conditional statement; if the statement returns false, the immediate next elif statement condition will be checked. If the elif statement returns false, the control jumps into the next elif statement and so on.

```
var = 100
if var == 200:
    print "1 - Got a true expression value"
    print var
elif var == 150:
```

```
    print "2 - Got a true expression value"
    print var
elif var == 100:
    print "3 - Got a true expression value"
    print var
else:
    print "4 - Got a false expression value"
    print var

print "Good bye!"
```

Another example of the greatest among three numbers is shown in the following code segment:

```
x = input("enter x")
y = input("enter y")
z = input("enter z")

if (x>y):
    if (x>z):
        print "highest is x:",x
    else:
        print "highest is z:",z
else:
    if (y>z):
        print "highest is y:",y
    else:
        print "highest is z:",z
```

12.2.2 Ternary Operator

The ternary operator is a useful operator that works similar to the `if-else` statement. It is also known as `if-then-else` operator. The syntax of the ternary operator is shown as follows:

```
a=5
b=6
var = a if (a>b) else b
```

If the value of a is greater than the value of b, the left-hand side data will be stored in var; if it is false, then the right-hand side data will be stored in var.

The more common example of greatest among three integer data can also be evaluated with ternary operator as follows:

```
a = input("enter a")
b = input("enter b")
c = input("enter c")
```

```
g = (a if(a>c)else c)if(a>b) else ( b if(b>c)else c);

print "highest is", g
```

12.3 Loops in Python

Loops are the essential component in a programming language. They are the code segments that perform repetition of tasks in a sequential manner based on the boundary value supplied by the loop. The loop boundary is the condition that mainly defines the termination of the loop control. In Python, we generally use while, for, and do-while loops (Figure 12.2).

12.3.1 while Loop

Like other programming languages, while loop consists of a boundary condition expression followed by a:symbol. The loop block will be executed based on the boundary value, and the counter will be accommodated inside the loop.

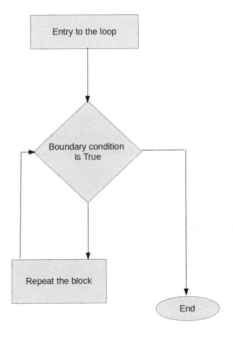

FIGURE 12.2
Loop structure in Python.

```
count = 0
while (count < 9):
    print 'The count value is:', count
    count = count + 1

print " loop completed"
```

In this case, while() evaluates the result based on the nonzero or true value. If the value of the condition is nonzero or true, the loop will be iterated, and if it is zero or false, then it will be terminated immediately. The following lines in Python will create infinite loops:

```
var =1
while(var):
        print "hi",var


var=True
while(var)
        print "hii", var
```

Sometimes, we do use while loop as an odd loop. Doing so, we can define the boundary of the loop during runtime. The following code takes the character input as the boundary of the loop and checks the value in the while loop in each iteration. As the value matches the loop, it will continue; otherwise, it will terminate.

```
count = 'a'
while (count != 'n'):
    print 'The count is:', count
    count = raw_input("enter count")


print "loop end"
```

And the output will be

```
The count is: a
enter count
The count is: e
enter count
end loop
>>>
```

Python also supports the loop statement associated with an else statement. We can use the else statement with while loop. If the condition of while loop is false, then the else statement will be executed. In case of for loop, the else statement will be executed when the loop gets exhausted. The following code illustrates the while loop with an else statement:

```
c = 0
while c < 10:
    print count, " the value is", c
    count = count + 1
else:
    print  " condition is false"
```

do-while loop is not allowed in Python, but we can emulate the feature of do-while using while loop. The following code illustrates the representation of do-while using while loop:

```
i = 1

while True:
    print(i)
    i = i + 1
    if(i > 4):
        break
```

12.3.2 for Loop

for loop is primarily used for iterating a list of values. It is mostly useful for iterating over list, tuple, and dictionaries. If the sequence contains any list, first of all, it will evaluate, and then the first item gets assigned to an iteration variable. Next, the statement block gets executed. The iteration procedure will run until the list gets exhausted. The following code segment illustrates an example of for loop:

```
for l in 'MyPython':      #  Example 1
    print 'Current Letter :',  l
car = ['ford', 'audi',   'bmw']
for c in car:         # Example 2
    print 'Current car :',  c
print " end for"
```

Another alternative approach to print the list using for loop is to iterate by applying the index of each element present in the loop. In such case, the index will start from 0 and end at size − 1.

```
car = ['audi', 'bmw',   'ford']
for i in range(len(car)):
    print 'Current car :',  car[i], i
```

And the output will be

```
Current car: audi 0
Current car: bmw 1
Current car: ford 2
```

The following is the prime number example:

```
for num in range(lower,upper + 1):
    for i in range(2,num):
        if (num % i) == 0:
            break
    else:
        print '%d is a prime number \n' % (num),
```

12.3.3 Nested Loops

In various programs, we often use nested loops. It is the property through which a loop can be accommodated in another loop [3]. One or more loops at a time can be placed in a nested loop. The number of loops nested in a loop determines the nesting depth of the loop. The following program shows the available prime numbers between 2 and 100:

```
i = 2
while(i < 100):
    j = 2
    while(j <= (i/j)):
        if not(i%j): break
        j = j + 1
    if (j > i/j) : print i, " is prime"
    i = i + 1
```

12.3.4 Odd Loops

Odd loops are the loop construction in which the boundary condition of the loop is supplied during the time of iteration of the loop. Therefore, we perhaps don't know when the loop will exactly end. It completely depends upon the user that when he/she wants to exit from the specified loop. The construction of the odd loop is given as follows:

```
ch = 'y'
while (ch <> n):
        print " the value of ch = ", ch
        ch = raw_input('enter ch')
```

12.3.5 break and continue Statements

The break statement is mainly used to break the execution of the current loop, and the control goes to the next instruction in this case. It is mostly used in both while and for loops. If we use break in the innermost loop, the inner loop will be broken.

```
l="MyPython"
for l in 'MyPython':
    if l == 'h':
        break
    print 'Current Letter:', l
```

On the other hand, continue statement immediately continues the iteration of the loop to the next pass by bypassing all the remaining statements after continue. This statement can also be used in both for and while loops.

```
for l in 'MyPython':
    if l == 'h':
        continue
    print 'Current Letter:', l
```

In this case, the letter h will be skipped by the loop and the rest of the letters will get printed normally.

12.4 Concepts Covered in This Chapter

- Decision-making
- Ternary operators
- Looping
- Odd loops
- Loop statements

References

1. Ellis, Margaret A., and Bjarne Stroustrup. *The Annotated C++ Reference Manual*. Addison-Wesley, Boston, MA, 1990.
2. Chun, Wesley. *Core Python Programming*. Vol. 1. Prentice Hall Professional, Upper Saddle River, NJ, 2001.
3. Zelle, John M. *Python Programming: An Introduction to Computer Science*. Beedle & Associates, Inc., Franklin, TN, 2004.

13

Functions in Python

Functions and the modules are the essential component of the programming language. A function is defined by a block of code that can be used repeatedly in a single program or in multiple applications too. Function ensures code reusability and modularity in a program. Using functions, a long piece of code generally segmentizes into a relatively tiny piece of module such that the program segment is easily readable. Some functions are user defined, while some are library functions. In case of Python language, the function declaration and definition mechanism has some specific technique. In this chapter, we discuss the different aspects and effectiveness of function in Python.

13.1 Standard Rule of Definition

There are some standard notations we have to maintain to define functions [1,2]. The function definition should start with "def" keyword followed by a name and opening and closing parentheses (name ()). The parentheses should contain any number of input parameters. The code block for a particular function should start with a ":" (colon) symbol. The function may contain return statements, and it returns one or multiple statements (by virtue of list properties). Functions have no return type, which is similar to return 0. A sample template or function definition is presented next. As the function parameter supports the positional order, you must mention the same order in which they are defined.

```
def function_name( function_parameters ):
    "function_docstring"
    The code associated in function

    return [expression]
```

The interesting thing here is that, in the function definition, we should pass only the name of the parameter but not the type. The following simple code describes a function to print a string message:

```
def print_msg(str):
        print str
        return
```

Sometimes, we can add a description or documentation string at the begin-ning of the function code segment. The following code segments reflect that fact:

```
def print_msg(str):
 " printing a string"
 print str
 return
```

13.2 Function Invocation

13.2.1 Invocation Methodology

After defining the function, we need to call it. The invocation of the function involves several types of techniques. We can directly call a function outside it, or sometimes, we can also call the function from another function too. The following code segment illustrates the technique:

```
# Function definition
def print_test( a,b ):
   "This prints the sum of two numbers"
   print "value is",a+b;
   return;

# invocation of the function
print_test(5,6)
```

In Python, the parameter passing in functions mainly uses the pass by refer-ence concept. It means the change made by the functions to the value actu-ally gives an impact to the original value. In Python, there is no concept of pointer, so the addressing operation has been done implicitly.

```
def change_list( list1 ):
   "This changes a passed list into this function";
   list1.append([10,20,30,40]);
   print "Values inside the function: ", list1;
   return;

#  call to list1 function
List1 = [1, 2, 3];
change_list( list1 );
```

```
print "Values outside the function: ", list1
```

In this case, the list object has been passed by reference and, in the function change_list(), the value of the content of the list gets appended with some new list values.

13.2.2 Argument Passing Mechanisms

In Python, two types of argument passing are possible: pass by values and pass by reference [3]. In case of the pass by value concept, only a copy of the values gets passed. All primitive types support the pass by value mechanism. On the other hand, object types always support pass by reference. If we want to pass list, tuples, dictionary, and string, it always follows the pass by reference mechanism. The following programs give an idea about the pass by value and pass by reference mechanisms:

```
#!/usr/bin/python


def swap(a,b):
     t=a
     a=b
     b=t
     print "values of a,b are",a,b;
     return

a1 = int(input("Enter a>"))
b1 = int(input("Enter b>"))

swap(a1,b1)

print "values a,b",a1,b1;
```

The output is

```
Enter a>5
Enter b>6
values of a,b are 6 5
values a,b 5 6
```

In this case, the values of a1 and b1 are passed to the swap() function and are received by the formal parameters named a and b. Then, the values are swapped using a third variable and get displayed in the same function. As a result, we get swapped values. But when we try to print the value from outside the swap() function, the value remains unchanged. It proves that the pass by value mechanism passes only the copy of the value of the variable.

In case of the pass by reference mechanism, the address or reference of the data object gets passed as follows:

```
#!/usr/bin/python

list1 = [10,20,30]

def pbr(list2):
      list2[1]=50
      print "list values", list2
      return;

pbr(list1)

print "list after passing", list1
```

The output is

```
list values [10, 50, 30]
list after passing [10, 50, 30]
```

Here, list is an object, so definitely the reference of the list will get passed through the function but not the copy of the list itself. Therefore, any change made to the actual value of the list will affect the original list of objects, and that change is available throughout the program.

13.3 Keyword Argument Concept

It is a very powerful concept in Python. If the parameters in the function parameter list are out of order, we use some keywords to identify them by their names. So if the order of the parameters does not match, the interpreter identifies the data object by their names. The following code shows the use of keyword arguments. The example shows that in the function calling part, the integer value is argument 1 and string is argument 2. But in the definition part, the order is reversed. Once the keyword is given to the function, the interpreter takes it and generates a proper function call.

```
def printfn( name, age ):
    "This function uses keyword argument"
    print "Name is: ", name;
    print "Age is ", age;
    return;
#call to printfn
printfn( age=32, name="piu" )
```

The default argument is another concept in which we do not give an argument during the call of the function. If, in the formal parameter, we place an explicit declaration of an argument, then that argument will be taken by default as the time of invoking. For example, we use two different calls for a single function in which one of the calls has one argument and the second one has two arguments. In the definition part, we put two arguments that are initialized explicitly.

```
#!/usr/bin/python
```

```
def info( name, age = 20 ):
    print "Name is: ", name;
    print "Age ", age;
    return;

# two different call of info
info( age=32, name="piu" )
info( name="piu" )
```

In the second case, the `info()` function takes the default value of age as 20.

13.4 Lambda Function

In Python, `lambda` keyword is primarily used to build an anonymous function often called lambda function. This function essentially does not have predefined names. This mechanism is suitable for creating adaptable functions and is therefore good for event handling.

The basic syntax for `lambda` function is as follows:

```
fn1 = lambda i: i**2
print(fn1(5))
```

Here, `lambda` function performs to the power (x^y) of the target input data `i`. A `lambda` function can also take more than one data as input.

```
fn2 = lambda a,b: a+b
print(fn2(3,6))
```

Another good use of `lambda` function is that if we want to generate an anonymous function at runtime, we can invoke it. Doing so at runtime, `lambda` returns the address of the function.

```
def fn3(n):
    return lambda i: i*n
```

```
dbl = fn3(2)
tri = fn3(3)
val = 10
print("Doubled value is: " + str(dbl(val)) + ". Tripled value
is: " + str(tri(val)))
```

13.5 Modules in Python

The module in Python is a logical concept that organizes the code logically. Using this approach, we can group the related codes in the same umbrella. It consists of an arbitrary name–value attribute. In Python, a simple module can be made and invoked through the name of the Python file itself. The module consists of function classes and different variables.

```
#Test.py
def print_val( var ):
    print "Hiii : ", var
    return;
```

The preceding piece of the code snippet basically acts as a module as it is imported by another file. The `Import` statement adds any source file with another file. As we import any file, the interpreter searches the path to the current directory for the availability of the file. If it is not found, then the search will be done in that path that has been written in the PYTHONPATH environment. The filesys.path stores the PYTHONPATH which is in /usr/local/lib/python in UNIX and C:\python22\lib in Windows

```
#importing the test module
Import Test
Test.print_val(5);
```

13.5.1 `from` Statement

`from` statement specifically imports some of the attributes from one Python module. The entire module, in that case, does not load, and only the selective attributes will load. Multiple attributes can be imported by maintaining comma (,).

```
from math import pi
from math import *
```

This is also possible to import all attributes from a module into the current namespace. This provides an easy technique to import all items from a module. When we attach any module to a source code, the Python interpreter searches the location of the module in the current directory first, and then it searches the directory maintained by PYTHONPATH. If it fails here, then it searched the default Python library. In case of Linux, it is /usr/local/lib/python/.

13.5.2 `dir()` Function

The built-in function `dir()` returns a list of attributes defined by the modules that are available. This helps us to identify the necessary attributes corresponding to a particular module in a sorted form.

```
import math
val = dir(math)
print val
```

The output is shown as follows (the sequence of attributes in the form of list):

```
['__doc__', '__name__', '__package__', 'acos', 'acosh',
'asin', 'asinh', 'atan', 'atan2', 'atanh', 'ceil', 'copysign',
'cos', 'cosh', 'degrees', 'e', 'erf', 'erfc', 'exp', 'expm1',
'fabs', 'factorial', 'floor', 'fmod', 'frexp', 'fsum',
'gamma', 'hypot', 'isinf', 'isnan', 'ldexp', 'lgamma', 'log',
'log10', 'log1p', 'modf', 'pi', 'pow', 'radians', 'sin',
'sinh', 'sqrt', 'tan', 'tanh', 'trunc']
```

13.6 Package in Python

The package is a set of Python modules clubbed together in a single directory hierarchy. This environment may contain *n* number of packages as well as subpackages and so on.

In case of package, we place many Python files in a single directory. This directory is basically used as a Python package. To invoke all files from that directory, we have to write another file named __init__.py. To make all functions available for a particular directory, we have to put explicit import statement in this __init__.py. For example, if we have a directory named mymath and we put add.py, sub.py, and mult.py in this mymath directory.

Let's assume that add.py has a function add(), sub.py has a function sub(), and mult.py has a function mult(). So, to invoke all these features in __init__.py, we have to write the following steps:

```
from add import add
from sub import sub
from mult import mult
```

After doing this, we can import our mymath package to any other function by writing

```
import mymath
```

13.7 `reload()`, `global()`, and `local()`

When a script imports a module, the top portion executes only once. So if we want to execute multiple times, we use the `reload` function.

```
reload(module)
```

`globals()` and `locals()` return the global and local namespaces in the environment where the current module is running.

If `global()` is called, it returns those names that can be assessed globally. If `local()` is called, it returns those names that are accessed locally.

13.8 Concepts Covered in This Chapter

- Creation of function
- Creation of module
- Anonymous functions
- Packages and their utilities

References

1. McKellar, Jessica. *Introduction to Python*. O'Reilly Media, Inc., Sebastopol, CA, 2014.
2. Oliphant, Travis E. "Python for scientific computing." *Computing in Science & Engineering* 9, no. 3 (2007): 10–20.
3. Van Rossum, Guido. "More control flow tools." In *An Introduction to Python*. Edited by Fred L. Drake. Network Theory Ltd., Bristol, 2003.

14

More Examples of Modules and Functions with APIs

Function oriented aspect of the programming language gives a potent to segmentized, modularized and perform reusability of the program code. In case of a real life application scenario, we mostly focus on the modular approach of software design. The effectiveness of the modules in programming is that they drastically reduce the effort to find out errors. As a result an effortless application design is obvious. As the program is sliced into several components, searching and finding an error is quite easier. In continuation of Chapter 13, describe and design and use of numerous functions modules and API to illustrates the effectiveness of the code.

14.1 Accessing CSV Files

CSV stands for comma-separated value. CSV files are mostly large files having a large number of data often called data objects [1,2]. They are a simplified form of schema-less data. Compared with JavaScript Object Notation (JSON), CSV is more readable than JSON which strictly follows the scripting phenomena.

The data inside CSV comprises rows, and the elements separated by a comma (,) reflect columns. Every line in the CSV file is a row in the spreadsheet, whereas commas are used to define and separate cells in the spreadsheet.

CSV module comprises some of the built-infunctions, which are as follows:

- `csv.reader`
- `csv.writer`
- `csv.register_dialect`
- `csv.unregister_dialect`
- `csv.get_dialect`
- `csv.list_dialects`
- `csv.field_size_limit`

14.1.1 Reading CSV Data

To get the CSV data, we have to use a `reader()` function that generates reader objects. `reader()` function takes each line of files and makes list of all columns. Then, one can choose the columns to display. The following example illustrates the reading technique using the reader object:

```
import csv
import sys

f = open(sys.argv[1], 'rb')
read = csv.reader(f)
for row in read
print row

f.close()
```

In the preceding code, we have imported `csv` and `sys` packages. The `csv` package is the standard package for accessing CSV files. `sys` is a standard system package. `sys.argv` is the command-line argument that passes during runtime. In this case, the name of the filename will be `argv 1`, because `argv 0` is the name of the Python file itself. After doing this, we are using the `open` function to open the file object. Then, a reader object is created using `csv.reader(f)`. The `for` loop iterates the reader object that comprises rows, which get printed at the output. Finally, we should close the CSV file reader object.

14.1.2 Loading CSV Using Pandas

Pandas (Python for data analysis) is a powerful tool for data analysis. It is built on the NumPy, which is another powerful application that creates and visualizes the multidimensional data. The fundamental objective of Pandas is to create multidimensional data frames.

We can actually create data frames in different format data such as CSV, JSON, and Python dictionary. Once you load the data frame, you can shape the data into several valuable information for data analysis.

In this case, we have to import Pandas as `pd` and NumPy. Then, we have to call the `read_csv()` function with the parameter of the .csv file name. To see the information available in the CSV file, we can invoke the `head()` method of the msft object. Figures 14.1 and 14.2 illustrate the code and output, respectively.

The output for the code is shown in Figure 14.2.

14.2 Parsing JSON Data

JSON stands for JavaScript Object Notation. It is very popular among developers for serializing data. It is used for most of the web Application

FIGURE 14.1
The code for accessing CSV.

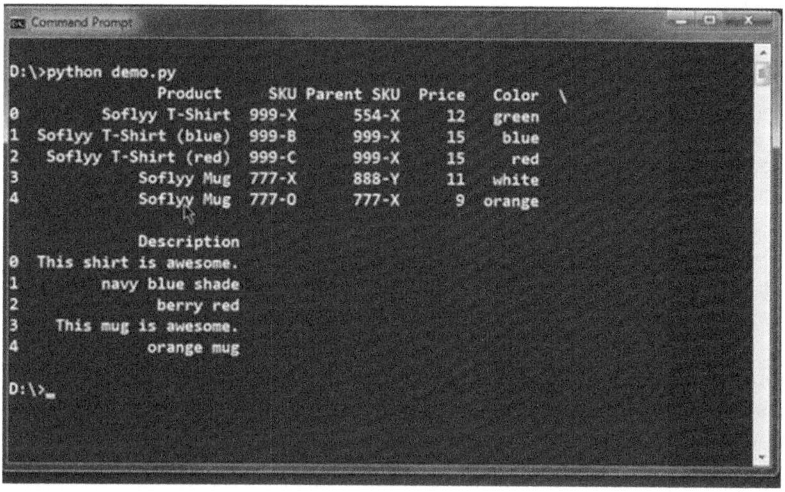

FIGURE 14.2
The output displayed by the head() method.

Programming Interface (web API) and passing of data between programs. Before accessing and parsing JSON data file, we have to import JSON module in our program.

We can create a JSON data file by simply making a dictionary-like data structure. In the following example, the basic framework of a JSON file is given.

```
J_data_1 = '{"a": 1,"b": 2,"c": 3,"d": 4}'
```

The standard JSON file uses the extension .json. A JSON example taken from json.org is given as follows:

```
{"widget": {
    "debug": "on",
    "window": {
```

```
            "title": "Sample Widget",
            "name": "main_window",
            "width": 550,
            "height": 550
        },
    "image": {
            "src": "Images/Sun.png",
            "name": "sun1",
            "hOffset": 270,
            "vOffset": 270,
            "alignment": "center"
        },
    "text": {
            "data": "Click to go",
            "size": 36,
            "style": "bold",
            "name": "text1",
            "hOffset": 260,
            "vOffset": 110,
            "alignment": "center",
            "onMouseUp": "sun1.opacity = (sun1.opacity / 100) *
90;"
        }
    }}
```

Although it has a dictionary-like structure, Python treats it as a set of strings unless it comes from a file. We can simply load it using the json.loads() method.

```
print json.loads(J_data)
```

It can be represented in a usable format. We can print the output using a loop too.

```
l_json = json.loads(J_data_1)
for m in l_json:
        print("%s: %d" % (m, l_json[m]))
```

14.3 Working with MongoDB

MongoDB is a free open source, cross-platform database system [3]. It is a document-oriented database program that is classified as a NoSQL database. It provides database as a service. A data server, in this case, has a collection of databases and comprises *n* number of data files. An important

MongoDB   Pymongo API   Python Script

FIGURE 14.3
Pymongo API.

advantage of MongoDB is that it supports the document-oriented approach; therefore, data is stored in the form of JSON-like object. It has schema-less design and is also easy to scale. The Python API for MongoDB is also known as PyMongo. The interaction between the database and the Python script through pymongo is shown in Figure 14.3.

We can download MongoDB from the following link directly: https://www.mongodb.org/dl/win32/i386. This link comprises several versions of 32-bit MongoDB installer file. Currently, MongoDB is also available in cloud version. Before installing MongoDB, we have to check the proper version that is compatible with our system. We can check it by using the following commands:

```
C:\>wmic os get osarchitecture
OSArchitecture
32-bit
```

We can directly install MongoDB from the suitable. MSI or .exe file.

To access MongoDB from Python, we should use pymongo API. You can install pymongo API using the following command:

```
$python -m pip install pymongo
```

We can update the existing one by applying the upgrade parameter.

```
$python -m pip installs –upgrade pymongo
```

Sometimes, we prefer to install it from the source. In that case, preferably we do install it from GitHub.

```
$ git clone git://github.com/mongodb/mongo-python-driver.git
pymongo
$ cd pymongo/
$ python setup.py install
```

MongoDB instance gets started by giving $mongod or sudo service mongod start command.

The first step is to make a connection with the mongod instance using MongoClient. It will connect the host and the port by default.

```
>>> from pymongo import MongoClient
>>> client = MongoClient()
```

We can also connect by supplying the hostname and port explicitly as follows:

```
>>> client = MongoClient('localhost', 27017)
>>> client = MongoClient('mongodb://localhost:27017/')
```

A single mongo instance supports multiple databases using attribute-style access or dictionary-style access as follows:

```
>>> coll = db.test_coll
>>> coll = db['test-coll']
```

Data in MongoDB is similar to the JSON script. Pymongo API actually uses a dictionary–like structure to represent the data. A sample data can be represented in the following way:

```
>>> import datetime
>>> ins  = {"author": "Amu Neel",
...          "text": "A Test Database!",
...          "tags": ["mongodb", "python", "pymongo", "API"],
...          "date": datetime.datetime.utcnow()}
```

To insert a date element into the MongoDB collection, generally insert_ one() method can be used. When a document is created, a default key (_id) is created by default. The value of the id is a unique key across the collection. As we insert an instance using insert_one(), it will immediately return an instance of InsertOneResult.

```
>>> posts = db.posts
>>> post_id = posts.insert_one(ins).inserted_id
>>> post_id
ObjectId('...')
```

To retrieve data, the most fundamental query that has to be implemented is find_one(). This method returns a single document, and it is very useful when the number of matching documents is only one.

```
>>> import pprint
>>> pprint.pprint(posts.find_one())
{u'_id': ObjectId('...'),
 u'author': u'Amu Neel',
```

```
u'date': datetime.datetime(...),
u'tags': [u'mongodb', u'python', u'pymongo',u'API'],
u'text': u'a Test Database!'}
```

14.4 MongoDB Client Management Tools

Robo 3T or Robomongo (Figure 14.4) is a cross-platform open source MongoDB management tool [3,4]. It embeds the same JavaScript engine that is used by the mongo management shell of MongoDB. This interface is available for Windows, Linux, and MacOSX. It is a free software, and the source is accessible through GitHub. We can download it from the following link: https://robomongo.org/.

The interface shows the connection that comprises logs, events, and stores. Data in the form of key and value can be visible in the middle of the interface. It is also possible to send the query string directly from the interface.

For MongoDB 4.X, a support for NoSQLBooster is also available (Figure 14.5). This is a shell-centric cross-platform interface that supports MongoDB2.6 version as well. This interface provides a very fluent query builder. The IDE itself is smart enough and has a built-in language service that knows all possible methods, properties, variables, and keywords. Besides, it also has the knowledge of MongoDB Field collection names. The new version is available in the following link: https://nosqlbooster.com/.

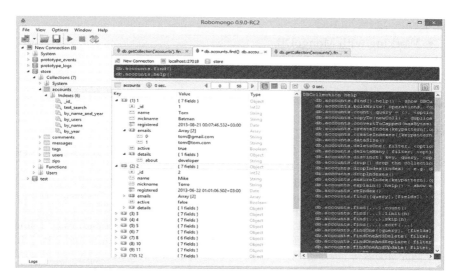

FIGURE 14.4
Robomongo interface.

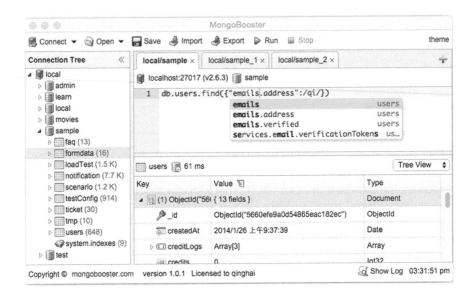

FIGURE 14.5
NoSQLBooster for MongoDB.

14.5 Concepts Covered in This Chapter

- Some of the usable and important APIs
- Accessing CSV data
- Parsing JSON files
- Working with MongoDB

References

1. Hellmann, Doug. *The Python Standard Library by Example.* Addison-Wesley, Boston, MA, 2011.
2. Herger, Lorraine M., and Mercy Bodarky. "Engaging students with open source technologies and Arduino." In *Integrated STEM Education Conference (ISEC), 2015 IEEE*, pp. 27–32. IEEE, 2015.
3. Kanoje, Sumitkumar, Varsha Powar, and Debajyoti Mukhopadhyay. "Using MongoDB for social networking website deciphering the pros and cons." In *Innovations in Information, Embedded and Communication Systems (ICIIECS), 2015 International Conference on*, pp. 1–3. IEEE, 2015.
4. Orr, Genevieve. "Computational thinking through programming and algorithmic art." In *SIGGRAPH 2009: Talks*, p. 31. ACM, 2009.

15

Implementation of Data Structures

The Data structure mainly emphasize on the fundamental organization of the data as data storage and data retrieval. It defines the effective storage mechanism of data elements in a systematic form. In programming language, data structure can be implemented with its inherent properties as well as by using third party APIs. Data structures can be categorized based on the ease of use. In C and other languages, the fundamental data structure that is dealt with is an array. We can derive various other data structures from the array too. Some of the data structures can be implemented using dynamic memory allocation. In Python, some of the efficient data structures that have been implemented are predefined. In this chapter, we discuss some of the elementary data structures in Python.

15.1 Lists in Python

List is the fundamental sequential data structure available in Python [1]. Each element is assigned by some data members. The structure can be accessed by its index. The base index of the list should start with 0 with an increment of 1 for each element.

We can perform certain operations on the list or sequence-type data structure. The primary operations that can be performed are slicing, indexing, addition, multiplication, and membership checking.

List is the most common type of data structure in Python [2]. It consists of a sequence of a data member of the same or different type. It is also called an ordered collection of data. By default, list is a dynamic structure, which means we can add or delete members. It is mutable, which means that it dynamically changes the type and size of the data objects stored in the list. The data members of the list stored by using square brackets ([]) are separated by commas (,). The index of the list starts with 0 and ends in the total size of the list minus 1. The following piece of code shows the declaration of the list and accessing the elements of the list:

```
l1 = ['honda', 'bmw', 2005, 2010];
l2 = [9,8,7,6,5,4,3,2,1];
print "list1[2]: ", l1[2];
print "list2[1:5]: ", l2[1:5];
```

In the above-mentioned example, we use list 1 whose elements are string and number types. The string member can be declared by "or." List 1 contains multiple types of data members, whereas list 2 contains the same type of data members. 11[2] will print the data of index location 2, which is 2005. In the case of 12[1:5], the sequence of data members will get printed. In this case, a slicing operation is done, which looks similar to indexing but not so much flexible. Here, 12[1:5] is basically a sublist starting with index 1 up to the fifth element by leaving the fifth indexed element. Therefore, the output will be as follows:

```
list1[0]:   [2005]
list2[1:5]:   [8, 7, 6, 5]
```

In the preceding program, we notice that we do use a semicolon (;). In Python, the semicolon is not mandatory but optional.

15.1.1 Updating and Deleting List Elements

Update and deletion operations can be possible directly in a list. To do so, we simply assign the values of the list elements to the variables based on the index positions of those values. It is illustrated as follows:

```
list = ['aaa', 'bbb', 15, 20];
print "Value available at index 2 : "
print list[2]

list[2] = 25;
print "New value available at index 2 : "
print list[2]
```

This code will immediately change the value of the index location 2 which is 15 to the new value 25.

To remove an element from a list, we necessarily use the del keyword. As a result, the value gets deleted.

```
list = ['aaa', 'bbb', 15, 20];
print "Value available at index 2 : "
print list[2]

del list[2];
print "New value available at index 2 : "
print list[2]
```

It will dynamically delete the element and the memory space corresponding to the index location 2. Another way to delete the data element is to use remove() or pop() function for the list object. The function will perform a similar type of operation. We have to supply the element itself in remove()

function to remove or delete the data. In the case of pop(), we have to apply the index location.

```
list = ['aaa', 'bbb', 15, 20];
list.remove('bbb');
print list[0:]
list.pop(2);
print list[0:]
```

The output will be

```
['aaa', 15, 20]
['aaa', 15]
```

The remove() function removes the string "bbb" and the size of the list is reduced by 3. The pop() function runs and removes the value 20 (index 2). Finally, it will become a list of two elements.

15.1.2 Fundamental List Operation

Various operations can be done in a list. The characteristic of the list is some-how similar to strings; therefore, the + operator is used for concatenation of the list. The * operator is applicable to the list to repeat an element *n* number of times. Table 15.1 shows some of the operations that can be performed in a list.

15.2 Tuples in Python

In Python, tuple is a sequence of immutable objects. The fundamental prop-erty of a tuple is that the values in the tuple cannot be changed.

To declare and create a tuple, we generally use parentheses. The values are to be separated by commas. Sometimes, they can be declared within

TABLE 15.1

List Operations

Expression	Output	Operation
['a','b','c']+['d','e','f']	['a','b','c','d','e','f']	Concatenation
['hello']*4	['hello', 'hello', 'hello', 'hello']	Repetition
Len(['a','b','c']	3	Length
'c' in ['a','b','c']	True	Membership

double quotes. Some of the standard declarations of the tuple are shown as follows:

```
Tpl = ('abc','def',1,2)

tpl2 = "a","b","cc","DD"

tpl3 = ()

tpl4 = (5,)
```

The third example (tpl3) shows a tuple that has no data. tpl4 is a tuple that contains only one element. We have use a comma after a single variable in the tuple.

   We can access the tuple data similar to accessing the data of the list.

```
Tpl = ('abc','def',1,2)

print "value of Tpl[0] is",Tpl[0]
```

We cannot update the elements of the tuple. Hence, the following declaration causes a runtime error and will stop the execution:

```
Tpl[1] = 'xyz' # en error
```

We can delete the elements of tuple using the del command. Therefore, the following code deletes one element of the tuple:

```
del Tup[1]
```

All other operations that are applicable to the lists are also applicable to the tuple.

15.3 Dictionary Structure

Dictionary is an unordered collection of objects. Unlike a normal object that comprises values only, the dictionary must have some key value pairs. It is an optimized structure that is used to retrieve the values based on some known key elements. The fundamental declaration of the dictionary structure is shown as follows:

```
dict = {'name': 'piu','age':32, 'grade':80}

print "Name is ",dict['name']
print "Age is ",dict['age']
```

We can update the new attribute in the dictionary; in such case, the new attribute will simply be added to the existing dictionary entry. The following code shows this:

```
dict['school'] = "Saradamani school"
```

We can perform deletion of the individual element of the dictionary as well as the complete dictionary.

```
dict = {'name': 'piu','age':32, 'grade':80}

del dict['age']

dict.clear()

del dict
```

The first statement deletes the attribute age. The second statement dict. clear() clears all attributes of the dictionary. Their del dict operation simply deletes the entire dictionary structure.

All other operations that are applicable to lists are also applicable to the dictionary.

15.4 Sorting of Data Structures

Sorting of data elements in data structures has an important role. Sorting performs arranging of data elements in either ascending or descending order. In this section, we discuss the sorting of data elements for different data structures. The sort() method is actually defined in most of the data structures to sort the data. Another way to sort the data elements is to call the sorted() function. The sorting technique for a list is shown as follows:

```
List1 = [ 'abc','def',123,'wrg']
List1.sort()
print  'sorted list', List1
```

This will sort the entire list. First, the numerical value will be printed, and then the value of the string will be sorted in alphabetical order.

We can also do reverse order sorting of the same data structure by making reverse parameter as True. Therefore, the code looks like this:

```
List1 = [ 'abc','def',123,'wrg']
List1.sort(reverse = True)
print  'sorted list', List1
```

It performs sorting of data elements in descending order.

We can also perform customized sort in a list. For example, a list object may contain two values, and if we are willing to sort the list based on the second element of the list object, then we have to perform the following code:

```
# second element is chosen for sort
def takeSecond(elem):
    return elem[1]
data = [(2, 2), (3, 4), (4, 1), (1, 3)]
data.sort(key=takeSecond)
print('Sorted list:', data)
```

Here, data list is sorted based on the second element of each list object. The element position has to be supplied by using the key parameter. The takesecond() function returns elem[1], which means the first indexed element of the list object is nothing but the second element of the list.

Sorting of the dictionary is a major issue. A dictionary can be sorted based on the key as well as the value. We can sort the dictionary by both the parameters of the dictionary. The following code shows the fundamental sorting of dictionary elements based on their keys as well as values.

```
dict = {'piu':32,'Nilanjan':34,'amu':33,'rishi':24,'bunku':24,
'mom':37}

def sortbykey():
    dict1 = sorted(dict.items(), key = lambda t: t[0])
    print dict1

def sortbyval():
    dict1 = sorted(dict.items(), key = lambda t: t[1])
    print dict1
```

To sort the dictionary elements, we have to check whether the sorting has to be done based on key or value. To do sorting based on the key, we should pass the key as the zeroth indexed element, which is t[0]. To do sorting based on the value, we should pass the first indexed value which is the second element. We can use a lambda operator for sorting the dictionary elements. lambda is a special type of operator/function that is used to create a small and one-time anonymous object or operation. It can have any number of arguments but only one expression. It never contains any statement. The following example shows the operation of a lambda function:

```
s = lambda a, b : a+ b
print s(2, 3)
```

Here, lambda takes the value of a and b as 2 and 3, respectively, and performs the sum() function. The sum() function can therefore be considered as a function for adding two numerical values. The operation of `lambda` operator is somehow similar to the macro used in C language.

15.5 Date and Time in Python

In Python, we can handle time and intervals using several techniques. Python core module can interact with the native computer system date and time. In Python, the time interval is the basic unit of second since 12:00 AM, January 1, 1970. The most efficient time module is available in Python that is used to represent and convert different time expressions. The method of `time.time()` always returns the system time from the 12:00 AM, January 1, 1970 epoch.

The following code shows the operation:

```
import time;

ticks = time.time();
Print "the number of ticks", ticks;
```

The date before the epoch cannot be represented with this representation [3]. The date in the far future is also not represented with this format of representation. The cutoff point of the date is 2038 as supported by UNIX and Windows. Time tuple has a great significance in date manipulation. Most of the Python time functions represents the time in the form of nine tuples. The attribute of the time tuple is mentioned in Table 15.2.

From Table 15.2, parameter 5 deals with second along with some leap second values to adjust the Universal Time Coordinated (UTC) time with universal time. Generally, when the time difference between UTC and universal

TABLE 15.2

Attributes of the Time Structure

Index	Attributes	Values
0	Year	2008
1	Month	1–12
2	Day	1–31
3	Hour	0–23
4	Minute	0–59
5	Second	0–60/61 (considering leap seconds)
6	Day of week	0–6 (0 = Monday)
7	Day of year	1–366 (Julian Day)
8	Daylight saving	−1, 0, 1, −1 library determine Daylight Saving Time (DST)

time is greater than 0.9 s, a one second leap time will be added in that case. Parameter 8 defines the daylight saving value that adjusts the time of daylight based on the latitude of the different geographical location.

To get the current time, we may call the `localtime()` method and pass the `time.time()` value through it. The following code represents the time shown by the system:

```
#!/usr/bin/python
import time;

ltime = time.localtime(time.time())
print "Current Local  time :", ltime
```

The output will be an unformatted form of time, but we can get formatted time from this representation. We can do any type of format time using the `asctime()` function.

```
import time;

lltime = time.asctime( time.localtime(time.time()) )
print "Local current time :", lltime
```

We can get the calendar month using the `calendar` module. There are various calendar methods through which we can perform several calendar operations.

```
import calendar

c = calendar.month(2008, 1)
print "Here is the calendar:"
print c
```

15.6 Strings in Python

String in Python is basically represented by a set of characters in the form of literal surrounded by single or double quotes. Creating a string is just equivalent to taking the values in a variable. The following code performs this task:

```
S='hello'
S1="hello"
```

We can access the character set of the string by simply traversing it using its index location. Therefore, the following declaration is valid:

```
s="hello python"
```

```
Print "string is",s[1:5]
```

This prints the string from index 1 to location 5 that is index 4.

We can update the value of the existing string or sometimes the substring by applying a reassignment to the value of the string. The following code performs this work:

```
v= "abc def"
```

```
print "after update", s[:5]+'python'
```

The output will be

```
after update abc python
```

We can remove the white space of a string using the `strip()` method.

```
v= 'abc def'
print (V.strip())
```

The `len()` function returns the length of the string.

```
print(V.len())
```

The `lower()` function produces a lowercase string if any character in the string is uppercase. The `upper()` function performs just the reverse operation.

```
print(v.lower())
print(n.upper())
```

The `replace()` function replaces all letters with a targeted letter

```
print(v.replace("a","k"))
```

15.7 Concepts Covered in This Chapter

- Lists
- Tuples
- Dictionary
- Date and time and their manipulation
- String manipulation

References

1. Miller, Bradley N., and David L. Ranum. *Problem Solving with Algorithms and Data Structures Using Python*, Second Edition. Franklin, Beedle & Associates Inc., Portland, OR, 2011.
2. Reed, David M., and John M. Zelle. *Data Structures and Algorithms Using Python and C++*. Franklin, Beedle & Associates, Portland, OR, 2009.
3. Dubois, Paul F., Konrad Hinsen, and James Hugunin. "Numerical python." *Computers in Physics* 10, no. 3 (1996): 262–267.

16

Object-Oriented Programming in Python

One of the biggest phenomena of Python is that it supports multi-paradigm programming approaches. The object-oriented feature is one of the most powerful approaches in Python, which maps real-life scenarios much comfortably in the form of a programming language [1]. The fundamental building block of object-oriented programming is the object. The object may be anything that represents any real-life thing. Other vital features such as polymorphism and inheritance are also supported by Python language. The main objective of the object-oriented paradigm is to achieve a level of reusability, sometimes known as DRY (don't repeat yourself).

16.1 Class in Python

In a generic sense, class is an elementary unit of object-oriented programming [2,3]. The class is sometimes said to be the blueprint of a real-life thing. A class contains properties such as variables and methods or functions which are the operations that can be performed over those variables. The following example illustrates the creation of a class and an object of the class (Figure 16.1).

```
class OS:
    # class attribute
    fs = "ext4"
    # instance attribute
    def __init__(self, name, ver):
        self.name = name
        self.ver = ver
```

```
root@amartya-Aspire-4736Z:/home/amartya/Desktop/python# python clas2.py
redhat is a ext4
ubuntu is also a ext4
redhat is 17 version
ubuntu is 16 version
```

FIGURE 16.1
The output for the class OS.

```
# instantiate the Parrot class
r = OS("redhat", 17)
u = OS("ubuntu", 16)

# access the class attributes
print("redhat is a {}".format(r.__class__.fs))
print("ubuntu is also a {}".format(u.__class__.fs))

# access the instance attributes
print("{} is {} version ".format( r.name, r.ver))
print("{} is {} version ".format( u.name, u.ver))
```

In the preceding code, a class OS having a class attribute fs and two instance attributes name and ver is created. The class attribute fs is printed using r.__cls__.fs, and the instance attributes name and ver are printed using dot operator (.). def __init__() is an initialization method, often called the constructor.

16.2 Constructor

Constructor is a special method that is referred by the syntax __init__().
The main job of the constructor is to initialize the instance attribute. We can use both parameterized and nonparameterized constructors. The parameterized constructor is used to initialize the attribute by using a customized value.

```
class Person:
        # no parameter
        def __init__(self):
            print(" non parametrized constructor example")
        def show(self,name):
            print("Hello",name)
    p = Person()
    p.show("Eshita")
```

In the preceding code, a nonparameterized constructor is made. In the next case, a parameterized constructor is discussed.

```
class Person:
        #  Parameterized Constructor
        def __init__(self, name):
            print(" parametrized constructor")
            self.name = name
        def show(self):
```

```
        print ("Hello",self.name)
    p = Person ("Eshita")
    p.show()
```

16.3 Creation of Methods in Class

Method is an essential component of class. It is used to perform a certain task. It is also used to define the behavior of the object [4]. The method has one, two, or more than two parameters. Sometimes, there is no parameter method that can be built. In the following code, the constructor initializes the name and ver attributes, and then, the GUI() method with two parameters and the Audio() method with one parameter are called.

```
class Linux:
    # instance attributes
    def __init__(self, name, ver):
        self.name = name
        self.ver = ver

    # instance method
    def GUI(self, graphics):
        return "{} graphics {}".format(self.name, graphics)

    def Audio(self):
        return "{} the name ".format(self.name)

# instantiate the object
u = Linux("ubuntu", 16)
# call our instance methods
print(u.GUI(" HD "))
print(u.Audio())
```

16.4 Polymorphism in Python

Polymorphism is the most fundamental property of the object-oriented programming [5]. Python supports polymorphism in the form of method overloading. The following code example simply reflects the polymorphism of different methods having the same name in two different classes (Figure 16.2).

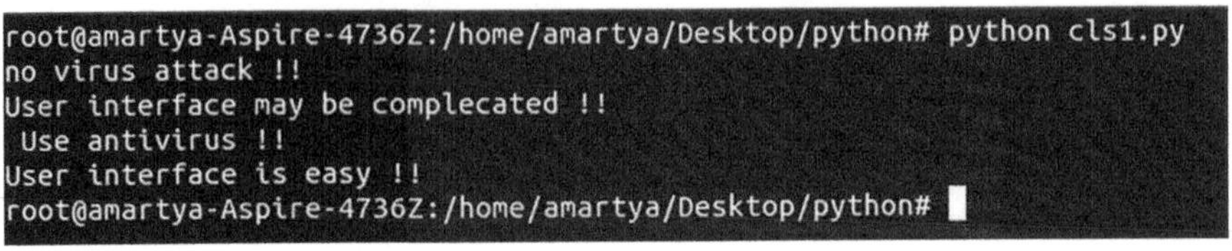

```
root@amartya-Aspire-4736Z:/home/amartya/Desktop/python# python cls1.py
no virus attack !!
User interface may be complecated !!
 Use antivirus !!
User interface is easy !!
root@amartya-Aspire-4736Z:/home/amartya/Desktop/python# █
```

FIGURE 16.2
Method polymorphism in the class.

```
class Linux:
    def virus(self):
        print("no virus attack !!")
    def ui(self):
        print("User interface may be completed !!")
class Windows:
    def virus(self):
        print(" Use antivirus !!")

    def ui(self):
        print("User interface is easy !!")

# a common interface
def os_test(os):
    os.virus()
    os.ui()
#instantiate objects
L = Linux()
W = Windows()
# passing the object
os_test(L)
os_test(W)
```

In the preceding code, there are two different classes: class Linux and class Windows. Both have the same virus () and ui() methods. A common interface is os_test () through which the object of both classes passes and finally the method will be invoked. Due to the polymorphism property in this case, different versions of the method are invoked based on the object that passes. This illustrates a simple demonstration of the polymorphic operation of the os_test () interface.

The polymorphism can be realized in Python by means of operator overloading. In Python, the "+"operator is already overloaded. This operator works for both "int" class and string class of the instance. Even we can overload an operator for some specific purposes and instances of specific classes. To overload the "+" operator from a custom object, we have to use the __add__ method.

```
import math
class Circle:
    def __init__(self, rad):
```

```
        self.__rad = rad
    def getRad(self):
        return self.__rad
    def area(self):
        return math.pi * self.__rad ** 2
    def __add__(self, a_circle):
        return Circle( self.__rad + a_circle.__rad )
c1 = Circle(4)
v1 = c1.getRad()
v2 = c1.area()
print 'radius of c1 > ', v1
print 'area of c1',v2
c2 = Circle(5)
v3 = c2.getRad()
v4 = c2.area()
print 'radius of c2' ,v3
print 'area of c2',v4
c3 = c1 + c2 # This became possible because we have overloaded
+ operator
                        # by adding a    method named __add__
v5 = c3.getRad()
print'radius of c3' ,v5
print 'area of c3',c3.area()
```

The output of the code is shown as follows:

```
root@amartya-Aspire-4736Z:/home/amartya/Desktop/python# python
overload.py
radius of c1 > 4
area of c1 50.2654824574
radius of c2 5
area of c2 78.5398163397
radius of c3 9
area of c3 254.469004941
```

16.5 Inheritance Concept

Inheritance is the most important phenomenon in the object-oriented programming [6]. This feature ensures reusability of the program code from one module or API to another module of the code. To realize this fact, we have to design the class hierarchy in which one class extends the properties of another class in a different manner. Various types of inheritance may exist in this case. Single inheritance ensures the extending properties from one class to another in a single level. This level can be extended further to achieve a multilevel of inheritance. The two most common types of inheritance are hierarchical and multilevel inheritances. In the hierarchical inheritance, more than one subclass inherits one base class, whereas in multiple inheritances one subclass inherits *n* number of base classes. The following program demonstrates the hierarchical inheritance feature, where a base class Unix is inherited by two subclasses: Linux and Android.

```python
class Unix:
    '''Represents Base of OS'''
    def __init__(self, name, ver):
        self.name = name
        self.ver = ver
        print('(Initialized os : {})'.format(self.name))

    def tell(self):
        '''Tell OS details.'''
        print('Name is:"{}" version:"{}"'.format(self.name,
self.ver),)

class Linux(Unix):
    '''Represents Linux OS.'''
    def __init__(self, name, ver, ui):
        Unix.__init__(self, name, ver)
        self.ui = ui
        print('(Initialized Linux: {})'.format(self.name))

    def tell(self):
        Unix.tell(self)
        print('Ui is: ',self.ui)

class Android(Unix):
    '''Represents Android OS'''
    def __init__(self, name, ver, v_name):
        Unix.__init__(self, name, ver)
        self.v_name = v_name
        print('(Initialized Unix: {})'.format(self.name))

    def tell(self):
        Unix.tell(self)
        print('Version name: ',self.v_name)

l = Linux('Ubuntu', 19, 'crystal')
a = Android('Android', 9 , 'Pie')

#  a blank
print()

os_list = [l, a]
for os in os_list:
    # Works for both operating systems
    os.tell()
```

The output looks as follows:

```
(Initialized os: Ubuntu)
(Initialized Linux: Ubuntu)
(Initialized os: Android)
(Initialized Unix: Android)
()
('Name is:"Ubuntu" version:"19"',)
('Ui is: ', 'crystal')
('Name is:"Android" version:"9"',)
('Version name: ', 'Pie')
```

In this case, Unix is the base class, and Linux and Android are the subclasses that inherit the properties of the base class Unix. Here, the __init__ () constructor initializes the members of the class as self.name = name and self.ver = ver. The tell () method describes the detailed feature of the OS itself. Further, subclasses Linux and Android of UNIX are created. Here, the __init__ () constructor is called and, within the body of the constructor, the base class constructor is called to ensure proper inheritance hierarchy. Finally, the objects of the two base classes are made, and by using those two objects, the tell() method is invoked. This can be done by creating a list of objects named os_list and iterate and print within the os_list object.

16.6 Method Overriding Concept

Method overriding is another important concept in the context of inheritance. In a class hierarchy, when the methods of the base class and the subclass have the same name and same type signature, they are supposed to be overridden. In this case, the method of the base class is basically hidden by the method of the subclass. Hence, the subclass method becomes active. However, the base class method, therefore, is invoked by the method of the subclass itself. The following code illustrates the method overriding. Here, base class A has the p() method, which is similar to the subclass B's method. Finally, the instance of B is created, and the method is done using b (which is the instance of B). The line b.p() illustrates the same.

```
class A:
    def __init__(self):
        pass
    def p(self):
        print ('base class')

class B:
    def __init__(self):
            pass
```

```
    def p(self):
        print ('subclass')

b = B()
b.p()
```

The output in this case is

```
C:\Documents and Settings\Administrator\Desktop\my-python\inh.
py =
subclass
>>>
```

16.7 Concepts Covered in This Chapter

- Introduction to class
- Adding constructor
- Adding method in the class
- Method and operator overloading
- Inheritance examples
- Method overriding

References

1. learnpython.org. Retrieved from https://www.learnpython.org/.
2. Goldwasser, Michael H., and David Letscher. *Object-Oriented Programming in Python*. Prentice Hall, Upper Saddle River, NJ, 2008.
3. Leuthäuser, Max. *"Lecture Note on "Object-Oriented Programming with Python."* Technische Universität Dresden, Dresden, Germany. http://st.inf.tu-dresden.de/files/teaching/ws10/ps/Leuthaeuser_Ausarbeitung.pdf.
4. Koenka, Israel J., Jorge Sáiz, and Peter C. Hauser. "Instrumentino: An open-source modular Python framework for controlling Arduino based experimental instruments." *Computer Physics Communications* 185, no. 10 (2014): 2724–2729.
5. GeeksforGeeks. Retrieved from https://www.geeksforgeeks.org/.
6. Lutz, Mark. *Learning Python: Powerful Object-Oriented Programming*. O'Reilly Media, Inc., Sebastopol, CA, 2013.

17

Input and Output in Python

Input and Output mechanism is the most common practice to take input to an application and send the output to another application or users. In the programming language like Python, there are several different tools to manage input and output. There are several API that serve the purpose of interacting with other remote application, the users, and even other devices or nodes in a network [1,2]. This chapter demonstrates several input and output mechanisms for Python.

17.1 `input()` and `raw_input()` Functions

The two most popular ways to take the input are using the `input()` and `raw_input()` functions. If we want to take any integer or floating point data element, we have to use the `input()` function. In this case, data will automatically be converted into a specific type and get stored in the variable c. The following example illustrates that:

```
a = input("enter a ")
b = input("Enter b")
c=a+b
print "Your value is: ", c
```

The `raw_input()` function is specifically used for taking the strings and character set as input. In this case, the data are collected directly as a console input stream. Therefore, all data are considered as a set of characters, and hence, no type is applied to it. Even if we enter any numerical value, it will simply be considered as a Unicode character. The following code snippet describes the functionality of the `raw_input()` function:

```
a = raw_input("enter a ")
b = raw_input("Enter b")
c=a+b
print "Your value is: ", c
```

This piece of code simply concatenates two strings. If the values are characters, they are taken as a string and get merged.

Through the input() function, we can also give the list of data values. The following code block can do the same:

```
st= input("enter a list")
print "List is", st
Enter a list: [12,13,14,15]
List is : [12, 13, 14, 15]
```

17.2 File Input/Output

Till now, we are dealing with a simple console input–output mechanism. In most of the cases, Python code deals with file input–output systems. The file input/output is mostly done by creating the object of the file, and there are several functions that perform the input–output operation through the file.

Before reading the file, we have to open it. Python provides the open() function to do so. The syntax of the open() function is shown as follows:

```
File o = open (file_name,Access_mode,buffering)
```

- file name: It is a simple string that suggests the name of the file to be opened.
- Access mode: It is the type of access applied to the file. The most popular access modes are like, read, write, and append. There are various other parameters used in Python. By default, the access mode is always read.
- buffering: buffering is basically an integer value that specifies the size of buffer during the input–output operation. If buffering is 0, then there is no buffer during file I/O. If it is 1, then it takes a buffer size of that value during I/O. If it is a negative number, then it takes a value which is a system by default.

In Table 17.1, we list the different opening modes of files.

TABLE 17.1

File Operation Modes in Python

Sl.	Modes	Properties
1	R	This is a default mode that opens the file for reading only.
2	Rb	This is a read mode that opens the file in a binary format. The pointer of the file is placed at the beginning of the file.
3	r+	It is a read/write mode for a binary file. The pointer is placed in the beginning of the file.
4	rb+	It is also a read/write mode for a binary file.
5	W	It is a write mode if the file is not present, and it has to be created when the data has to be written. It overwrites the existing file.
6	Wb	It is a written binary mode. It overwrites the existing file.
7	w+	It is a read/write mode for a binary file. The pointer is placed in the beginning of the file.
8	wb+	It is the read–write mode for a binary file.
9	A	This mode appends the file with the existing content. The pointer always starts with the end of the file.
10	Ab	This is an append mode for a binary file. The same operation is done for binary files.
11	ab+	This is a read/append mode for a binary file.
12	a+	This is a read/append mode for a normal file.

17.3 Attributes of the File Object

There are certain attributes in the file object that deal with the proper handling of any file. The attributes are mentioned in the following code:

```
#!/usr/bin/python

f = open("my.txt", "wb")
print "Name of the file: ", f.name
print "Closed or not : ", f.closed
print "Opening mode: ", f.mode
print "Softspace flag : ", f.softspace
ant the output will become...
amartya@amartya-Aspire-4736Z:~$ python test.py
Name of the file:  my.txt
Closed or not :  False
Opening mode :  wb
Softspace flag :  0
```

The first parameter f.name simply returns the name of the file that is invoked. The f.closed parameter is a Boolean value that suggests whether the file is closed or not. If it is true, it is closed; otherwise, it is false. f.mode returns the opening mode of the file, and f.softspace mentions any soft buffer in the file which is 0 in this case.

17.4 `close()` Operation

The close() operation ina file closes the file pointer by flashing any unwritten data in the file. After close() is executed, it will allow any reading and writing operations.

In a normal case, Python by default closes a file, as the reference of the file object gets assigned to any other file.

The following example shows the operation of the close() function:

```
f= open("my.txt", "w+")
print "Name of the opened file is: ", f.name

# Close file
f.close()
```

17.5 Reading and Writing a File

The read() method is used for reading the text and binary data from an opened file. In this method, we must pass the number of characters or bytes to be read at a time. The read() method always starts reading from the beginning of the file; if we don't provide the number count, it will read up to the end of the file.

```
myf = open("test.txt", "r+")
str = myf.read(15); #15 elements
print "Read data are: ", str
# Close
myf.close()
the output: Read data are:
helloworldthisisaninputoutputexample
```

The write() method writes the string or binary data to an opened file. In this method, we have to pass the parameter string to be written. When a file is opened in the write mode, all the previous contents will be erased.

```
#!/usr/bin/python
# Open a file
myf = open("test.txt", "wb")
myf.write( "helloworldthisisawritebinaryexample");
myf.close()
```

17.6 `tell()` Method

Often, we do want to know the current position of the file. To do this, we have to use the `tell()` method. This method directly returns the current positions of the read and write pointers within a given file. We should not pass any parameter through the `tell()` method. The operation of the `tell()` method is demonstrated as follows:

```
f = open("test.txt", "rw+")
print "Name:   ", f.name
l = f.readline()
print "Read Data: %s" % (l)
# current position.
p = f.tell()
print "Position of the pointer: %d" % (p)
# Close file
f.close()
```

The output looks like the following:

```
root@amartya-Aspire-4736Z:/home/amartya/Desktop/python# python
tell.py
Name:   test.txt
Read Data: This is line 1
Position of the pointer: 15
```

17.7 `seek()` Method

The `seek()` method primarily takes the current position in an offset value. There is an argument called whence. The value of whence determines the offset value of the file position. This argument set defaults to zero. To change the offset value of the file position, we do change the whence value. If the value is 1, then the `seek()` happens with respect to the current position. If the value is 2, then the `seek()` happens with respect to the end position of the file. The `seek()` operation does not return any value. Sometimes, the file may open in to write, append, or 'a+' mode (often called append and update). In such cases, the update of the file will reset the seek pointer position.

```
f= open("test.txt", "rw+")
print "Name : ", f.name
l = f.readline()
print "Read : %s" % (l)
f.seek(0, 0)
l = f.readline()
print "Read2 : %s" % (l)
f.close()
```

The output looks like the following:

```
root@amartya-Aspire-4736Z:/home/amartya/Desktop/python# python
seek.py
Name :  test.txt
Read : This is line 1
Read2 : This is line 1
```

17.8 Command-Line Argument in Python

Command-line argument is an argument that is passed to the program code during runtime. It is supposed to be one of the input-taking mechanisms. In other programming languages, this argument is passed through the main method. In case of Python, we can take command-line argument through the sys module. The technique in this case is shown as follows:

```
import sys
a1 = sys.argv[0]
```

We should always keep in mind that the argument indexing will start from 0, because, when we pass the argument sequentially during runtime, it is stored in a sequential list of arguments called argument vector. Here, argv[] denotes that argument vector through which the argument should pass.

In case of Python, argv[0], that is, 0th argument, should be the name of the Python file. Then, argv[1], argv[2],... will subsequently give the actual arguments that are passed during runtime.

```
import sys
File = sys.argv[0]
args = sys.argv[1:]
count = len(args)
```

```
print "name is", File
print "arguments are" , args
print "length", count
```

In the first case, the output shows the name of the file; in the second case, it shows the argument given; and in the third case, it prints the number of argument.

We can also print the command-line arguments using a for loop. In this case, the for loop iterates through the list of arguments and prints one by one.

```
import sys
for x in sys.argv:
    print "Argument: ", x
```

17.9 Concepts Covered in This Chapter

- The file I/O basics
- File object creation
- Reading and writing a file
- tell() method
- seek() method
- Command-line arguments

References

1. Lutz, Mark. *Programming Python*. O'Reilly Media, Inc., Sebastopol, CA, 2001.
2. Van Rossum, G. Python Programming Language. In *USENIX Annual Technical Conference* (Vol. 41, p. 36) (June 2007).

18

Exception Handling in Python

In a standard Python application development scenario, the wrong syntax and semantics leads to exceptional condition in the program. This kind of error naturally causes during the time of execution of the program. When such error occurs, the program should take an appropriate safety measures so that the exception get handled to avoid the instantaneous program crash by the system. Exception handling mechanism is a convenient way to monitor and handle errors under certain special conditions. One can feel the need of exception handling to safeguard the program execution from the possible unknown behavior of the program code.

18.1 Syntax Error vs. Exception

Syntax error, often called parser error, is a type of error when parser gets a set of codes that violate the rule of a language [1]. This is the most common type of error in Python. Any mistake in proper syntax generates an error like the one shown in Figure 18.1. Such errors are also highlighted with an arrow so that one can easily detect the point where the error actually occurs. For example, if the colon: used at the end of the while loop is missing, the arrow points to the statement immediately after the while

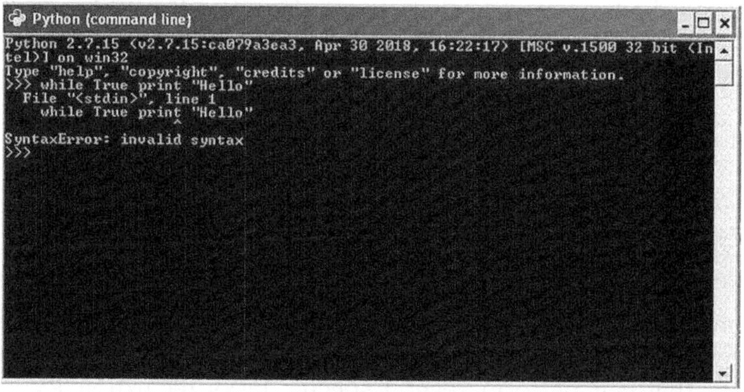

FIGURE 18.1
Error in Python console.

FIGURE 18.2
Error displayed in IDLE GUI.

loop. In case of IDLE Graphical User Interface, the error message is not displayed directly in the console; instead it shows a message dialog box as illustrated in Figure 18.2.

Suppose in a situation where a variable is divided by 0, as shown in the following code:

```
var  = 6

var2 = 0

print var/var2
```

Although the program is syntactically correct, the execution stops due to an extraordinary condition which is nothing but the ZeroDivisionError. The following message shows the file name along with the line number. Finally, it also shows the type of error that occurs. Several different types of exception such as ZeroDivisionError, TypeError, and NameError are available in Python library.

```
Traceback (most recent call last):
  File "C:/Python27/t2.py", line 7, in <module>
    print var/var2
ZeroDivisionError: integer division or modulo by zero
>>>
```

Exception that is generated in Python is actually the object of the Exception class. The base class of exception is Exception. There are several other subclasses that can perform several specific tasks regarding Exception. Table 18.1 shows the exception types available in Python, including their functionalities.

TABLE 18.1

Exception Types in Python

Sl.	Exception Types	Functionality
1	`Exception`	Base class of all exception types
2	`SystemExit`	Raised by the `System.exit()` function to exit the system
3	`StopIteration`	Raised by the `next()` method during iteration and when the iterator does not get any object further
4	`StandardException`	Base class of most of the standard built-in exception types
5	`ZeroDivisionError`	A special type of arithmetic error causes when a value is divided by zero
6	`ArithmeticError`	Base class of all errors related to numerical computation
7	`FloatingPointError`	Improper floating point calculation or calculation failure raised this exception
8	`OverFlowError`	Occurs when any computation exceeds the range of a certain numeric value
9	`EOFError`	Raised when the end of file is reached and also no input is returned by the `input()` or `raw_input()` function

18.2 Handling Exception

As the exception may be generated by a user program, we can even handle it using exception handlers. For example, we can make a code block that takes some integer data, and as soon as the data of any other format is given to it, it will immediately generate an exception [2].

18.2.1 Try-Except Technique

We use a `try: except:` block to handle exception. In this block, we can accommodate the code that we want to monitor in the `try:` block, followed by the `except:` block. If no exception occurs, then the `try:` block finishes its execution and the `except:` block gets skipped. When an exception occurs within the `try:` block, the control immediately stops the execution of the code and gets shifted to the `except:` block. Therefore, the code beyond the exceptional code will never be executed in this case. The following program shows the exceptional condition:

```
>>> while 1:
        try:
                x =int(input("data: "))
                print "an integer data"

        except:
                print "Wrong Format"
```

The output will be

```
data: 2
am integer data
data: t
Wrong Format
Data:
```

A single `try:` statement may have several exception clauses to specify the different issues for different types of exception. In this case, at least one handler must handle the corresponding exception. As the handler handles the exception, no other block can respond to the same exception. An exception clause sometimes holds multiple types of exception in the form of a tuple. The following code reflects the case. All the exceptions are basically compatible with the same class and the base class.

```
Except ( NameError, TypeError):
        print " Error !!"
```

In the following code, we describe how an exception class can be made by inheriting the base class exception. In this case, three classes have been made by inheriting the base class exception, and in the next part, the exception has been raised.

```
class X(Exception):
    pass

class Y(X):
    pass

class Z(Y):
    pass

for cls in [X, Y, Z]:
    try:
        raise cls()
    except Z:
        print("Z")
    except Y:
        print("Y")
    except X:
        print("X")
```

In this example, one thing to be noted is that we have to write the exceptions in reverse order, because if we do X, then Y, Z will also be caught by "except X" as the X is the parent class of Y and Z so the Exception will trigger the parent class and immediately execute it by bypassing all child classes.

18.2.2 Use of `else` Statement

Sometimes, the `try:except:` block may comprise an `else:`. In such case, if there is no exception raised in the program, the control will execute the `else:` part. In this case, the file name as an argument is opened by the program, and if an `OSError` opens that file, it will immediately treat it as an exception, and the message "unable to open" will be displayed. In another case, the character set gets extracted as we read the line and the length of the line gets printed.

```
for arg in sys.argv[1:]:
    try:
        f = open(arg, 'r')
    except OSError:
        print('unable to open', arg)
    else:
        print(arg, 'has', len(f.readlines()), 'lines')
        f.close()
```

In most cases, it happens that an except clause may specify a variable after the name of the exception. The variable in this case is bound to an exception instance with the arguments stored in `ins.args`. For easy understanding, the exception instance defines `__str__()`; therefore, the arguments can directly be printed without having to reference `.args`. We may also do instantiation of exception first before raising it and add any attributes to it as desired.

```
>>> try:
    raise Exception('Hi', 'Hello')
  except Exception as ins:
        print(type(ins))     # the exception instance
        print(ins.args)      # arguments stored in .args
        print(ins)           # __str__ allows args to be printed
directly,
                             # but may be overridden in exception
subclasses
        a, b = ins.args
        print('a =', a)
        print('b =', b)

<class 'Exception'>
('Hi', 'Hello')
('Hi', 'Hello')
x = Hi
y = Hello
```

Sometimes, exception may arise within a function instead of happening within an exception. In this situation, as the function invokes, the exception will be the immediate result. The following code illustrates the fact:

```
>>> def this_arithmetic():
      a = 14/0

>>> try:
        this_arithmetic()
    except ZeroDivisionError as er:
        print('Handling run-time  divided by zero arithmetic
error:', er)
```

Handling run-time divided by zero arithmetic error: division by zero.

It is also possible that we can raise an exceptional condition within the code itself. We can do it by using the raise keyword. The name of the exception can be passed through the raise ExceptionName().

```
>>> try:
        raise NameError('Demo')
    except NameError:
        print('An Exception occurred!!')
        raise
An Exception occurred!!
Traceback (most recent call last):
  File "<stdin>", line 2, in <module>
NameError: Demo
```

18.2.3 finally Block

The finally clause is another important clause used in all cases. The general rule for the finally clause is that the finally: block will always be executed irrespective of whether an exception occurs or not. The finally: block should be placed at the last part of the exception statement. When an exception occurs, it will be handled by the except statement or sometimes it will be handled by the else: statement. If the exception is successfully handled or not raised at all, the control will go to the finally: block.

```
>>> def div(a, b):
      try:
            res = a / b
        except ZeroDivisionError:
            print("error occurred !! division by zero!")
        else:
            print("result is as follows", res)
        finally:
            print("Execution of the finally block")

>>> div(2, 4)
result is 0.5
Execution of the finally block
```

```
>>> div(2, 0)
error occurred !! division by zero!
executing finally clause
>>> div("2", "4")
Execution of the finally block
Traceback (most recent call last):
  File "<stdin>", line 1, in <module>
  File "<stdin>", line 3, in divide
TypeError: unsupported operand type(s) for /: 'str' and 'str'
```

18.3 Concepts Covered in This Chapter

- Error and exception difference
- Handling mechanism
- Try-except block
- Use of `else`:
- Use of `finally`: block
- Multiple except statements

References

1. Lutz, Mark. *Programming Python: Powerful Object-Oriented Programming.* O'Reilly Media, Inc., Sebastopol, CA, 2010.
2. Hammond, Mark, and Andy Robinson. *Python Programming on Win32: Help for Windows Programmers.* O'Reilly Media, Inc., Sebastopol, CA, 2000.

19

Graphical User Interface in Python

Representation of software in a graphical user interface (GUI) is now common. Most of the modern programming platforms such as C++, Java, and Microsoft visual studio provide GUI components such as frames, forms, buttons, text box, combo box, and list box. The GUI makes the software more readable and interactive as well as more user-friendly. Nowadays, Android provides a more sophisticated and user-friendly GUI for smartphones and other smart gadgets. Python also has a huge number of GUI frameworks or toolkits in the form of libraries and APIs. The primary and classical GUI component in Python is Tkinter, which is bundled with Python using Tk. There are also some cross-platforms and native solutions available to build a platform-specific software.

19.1 Introduction to Tkinter

Tkinter is a GUI of a defacto standard [1,2]. It comprises an object-oriented layer on the top of Tcl/Tk. Tcl is a tool command language. It's very easy to learn and suitable for a wide range of programmers who deal with desktop applications, networking, and network administrative tasks.

Tkinter library is present by default in a Python library list that provides a fast and easy approach to develop GUI applications. Creation of GUI using Tkinter is a very easy job, and what you need to do is to maintain the following things:

1. Import Tkinter module at the beginning.
2. Create the main GUI application.
3. Add different widgets in the Windows application.
4. Enter into the main event loop and control the action of user-generated events.

In case of Windows, Tkinter package is available in the software bundle itself [3]. In case of Ubuntu Linux, we can install the package explicitly by applying sudo apt-get install python-tk or using the synaptic package manager (Figure 19.1).

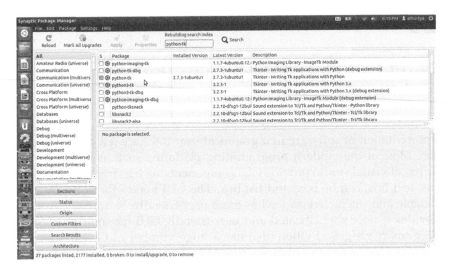

FIGURE 19.1
`python-tk` package installation.

After installation, we run the GUI. The code demonstrates a basic implementation of the Tkinter GUI:

```python
#!/usr/bin/python
import Tkinter
t = Tkinter.Tk()
# Code to add widgets will start.
t.mainloop()
```

Here, `Tkinter.Tk` returns the GUI components to `t.mainloop()`, which is the fundamental function that repeats the chunks of code that carry out the task under the application.

The output goes like this in Figure 19.2.

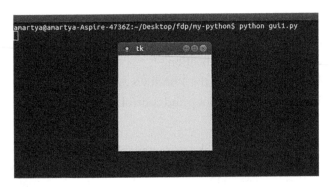

FIGURE 19.2
Basic Tkinter window.

Tkinter comprises several modules. Tk interface is supported by the binary extension of _tkinter that comprises the low-level interface of Tk. It is a shared library, and sometimes, a static-linked interface to Python interpreter is not used by the programmer directly.

Generally, Tkinter module exports several widget classes and their associated constants; therefore, we attach the module in all the ways as mentioned earlier (Figure 19.3).

```
import Tkinter

from Tkinter import *

import Tkinter as Tk

in the next program, we are going to print a hello message

from Tkinter import *

r = Tk()

l1 = Label(r, text="Hello,to Python Tkinter!")
l1.pack()

r.mainloop()
```

The program starts with Tkinter import * that imports all the classes and modules in the Tkinter toolkit package. In most of the cases, we should import all the modules in the modules namespace.

To initialize the base widget, we have to create r =Tk(). This is a simple window comprising the title bar and a default window decoration.

The Label() method is a widget that displays text or an image object in the default window. In our case, we show the "Hello, to Python Tkinter!" message. We call the pack() method to fit the size of the text into the default window, and, of course, it makes the text visible through the window.

FIGURE 19.3
Printing a message.

The next important method is the `mainloop()` method, often known as `r.mainloop()`. This is the main event loop and window that will never appear until this event loop is called. The code will run by the event loop until the close button is closed. The eventloop controls all the features related to an event such as mouse click and key pressed. Not only the event but also the event loop control the operation of the Tkinter itself.

19.2 Core Component Classes

There are some fundamental classes that are the building blocks of Tkinter. A total of 15 widgets are supported by Tkinter: button, canvas, checkbutton, entry, frame, label, listbox, menu, menubutton, message, radiobutton, scale, scrollbar, text, and Toplevel. In Python 2.3, there are some more widgets :LabelFrame, spinbox, and panedwindow. Tkinter module gives numerous classes in various widgets in tk and various mixins also. A mixin is a class design that blends with other classes through multiple inheritances.

There are three main classes that are used by mixing through widget classes: Grid, Pack, and Place.

Grid is a geometry manager that allows the programmer to create table-like layouts by organizing components in a two-dimensional grid. Pack is another efficient geometry manager that packs the widgets into another parent widget by providing a specific layout so that the arrangement becomes ordered. Place geometry manager lets the widget to be placed in an explicitly given position.

Before we go further, let us implement another simple example that illustrates the fundamental component of Tkinter.

```
from Tkinter import *

class App:

    def __init__(self, master):

        f = Frame(master)
        f.pack()

        self.button = Button(
            f, text="EXIT", fg="blue", command=f.quit
            )
        self.button.pack(side=LEFT)

        self.hi_there = Button(f, text="HII",
command=self.s_hi)
```

FIGURE 19.4
Windows with button widgets.

```
        self.hi_there.pack(side=LEFT)

    def s_hi(self):
        print "This is about button component!"

r = Tk()

app = App(r)

r.mainloop()
```

The preceding code produces an output as shown in Figure 19.4.

Here, in this case, the entire widget is written within a class. The constructor of the class (def __init__ ()) creates the frame first, and then within the frame, it creates two buttons. The first button named as "EXIT" button holds a quit event labeled as "HII" calls the method s_hi() and prints the message as it invokes by clicking the "HII" button. All buttons have been packed from the left side. Then, we create some script-level coding to create tk widgets. The App class instance will be created, and then main loop will be generated to send TK to the event loop until the "EXIT" button is pressed.

19.2.1 Button Widget

Button is a fundamental widget for any front-end application. It can be used to perform various tasks such as controlling operations, loading data from external resources, and monitoring certain tasks in realtime. It may be associated with a method, and we can link it to that method. The Tkinter automatically calls that method as the button is pressed. Each button in Tkinter has an inbuilt event listener that listens to the event and performs a task based on that event.

Buttons are of different types. Sometimes, we use radio and check buttons. There are various purposes we do serve through buttons such as ok, cancel, yes, or no button that is mostly used for the choice.

To create a button, we may mention the specific content and function which is under the button itself. A simple declaration of a button is shown as follows:

```
from Tkinter import *

m= Tk()

def callme():
     print "A click event!"

b = Button(m, text="OK", command=callme)
b.pack()

mainloop()
```

Here, m instantiates Tk. We define the callme() method that prints a message. Then, a button is created with a text OK. The callme() method is called under the click event of the button. Finally, it is packed, and the main loop is called. However, the state of the button can be changed as a disabled state by setting the state parameter as DISABLED.

```
B = Button (m , text = "hello" , state = DISABLED)
```

By default, the size of the button is enough to hold a standard size text, but we can use padx and pady to add some extra space in the button. We can also use height and width to set the height and width of the button. If the button consists of an image, then we can use pixel value instead of height and width values. The following code shows one of the ways to do that:

```
fr = Frame(m, height=42, width=42)
fr.pack_propagate(0)
fr.pack()

btn = Button(fr, text="Happy!")
btn.pack(fill=BOTH, expand=1)
```

We can add an anchor and justify parameters to set the position of the button too. The following code snippet shows the code for the same:

```
b = Button( m, text = bigtext, anchor = W, justify = right,
padex = 3)
```

A config() function can also be used to make the configuration of the button from raised to sunken. We can call it from the button instance b.config(relief = SUNKEN).

Some of the important attributes of the Button class are listed in Table 19.1.

TABLE 19.1

List of Attributes for Button

Attribute (Widget Option)	Phenomenon
Activebackground	The backgroud color that is used when the button is active. By default, it is system specific
Activeforeground	The foreground color during the active button
Anchor	Controls where the button text or image is located. Typically, N,S,E,W,NE,NW,SE,SW
Background	Background color. By default, it is system specific
Bg	Sets the background
Bitmap	The bitmap to display a widget
Borderwidth/bd	Defines the width of the border of the button. By default, it is 1–2 pixels
Cursor	To show the mouse pointer appearing over the button
Default	If it is set, the button acts as a default platform-specific button
Font	Font used in the button. Supports one type of font at a time
Foreground/fg	Specifies the foreground color of the font
Height	The height of the button itself. If it is omitted, the height is calculated based on the content of the button
Highlightcolor	Specifies the color of the button when it is focused. The default is system specific
Highlightthickness	The thickness of the highlighted border
Image	The image to display in the widgets
Justify	This aligns the button in the frame with LEFT, RIGHT, or CENTER parameter
Overrelief	Alternative to relief when the mouse is moving over the widgets
Relief	This is the border decoration wither sunken or raised. Some other possible values are groove flat
State	State of the button. Often marked as ACTIVE or DISABLED; the default is NORMAL
text	Text to display in the button. Sometimes, text, bitmaps, or image is also allowed
Underline	Character to underline in the text label. The default is no underline which is coded by –1
Width	The width of the button
Wraplength	Length of the text wrapped into multiple lines

19.2.2 Canvas Widget

Canvas is a popular widget that provides a complete graphical support for Python GUI design. This versatile widget is mainly used for drawing, plotting, and creating graphical editors as well. It is a typical widget used to display graphics and another drawing.

To draw the thing in canvas, we can use the create method (Figure 19.5).

FIGURE 19.5
Canvas widgets.

```
from Tkinter import *
m = Tk()
w = Canvas(m, width=200, height=100)
w.pack()
w.create_line(0, 0, 200, 100)
w.create_line(0, 100, 200, 0, fill="blue", dash=(2, 2))
w.create_rectangle(50, 25, 150, 75, fill="yellow")
mainloop()
```

19.2.3 CheckButton Widget

Checkbutton widget class is primarily used to create a button that mentions the on/off state. It is a standard Tkinter widget that contains a text or image, and one can associate any Python function with each of the check buttons. When a button is pressed, an event is generated, and hence, the call of the function is made. When we add a text corresponding to a button, it by default takes a single font. Having multiple lines for a button is allowed. By default, it also takes an underline to a specific character to show the keyboard shortcut for that particular button.

Primarily, the checkbox reflects an ON/OFF relation; however, multiple checkboxes endorse a many-to-many relation. Often, we create a one-to-many relation by means of the list and the radio button.

To use a checkbox, first we have to create a Tk variable and then a checkbox object.

```
from Tkinter import *
m = Tk()
v = IntVar()
def call1():
    print "test check box!!"
c = Checkbutton(m, text="Start", variable=v, command = call1)
c.pack()
mainloop()
```

In the preceding program, a checkbox is created, and the text is set to "Start." As we do check it, the command will call the call1() method.

As a result, it prints the content of the `print` statement. The variable `v` is by default 0, and when it is checked, the value gets toggled to 1. However, we can change the default value by the "onvalue"/"offvalue" option.

19.2.4 `Entry` Widget

The `entry` widgets primarily involve entering the `text` string. This widget allows the `text` to be entered in a single line. To add the `entry text` to the widget, the `insert()` method can be used. You can call the `delete()` method, before you insert a new `text` to replace the current `text`.

```
me = Entry(m)
me.pack()

me.delete(0, END)
me.insert(0, "a default value")
```

To fetch the entry text for current entry we have to call get method.

```
s = e.get()
```

Entry widget is bound by a `StringVar` instance and performs setting or getting the `entry text` through that variable.

```
var = StringVar()
ent = Entry(master, textvariable=var)
ent.pack()
var.set(" test default")
str = var.get()
```

The following program describes the technique that takes some `text` through the `entry` widget, and by pressing a `button`, the current `text` is displayed through the `entry` widget.

```
from Tkinter import *

m = Tk()
e = Entry(m)
e.pack()
e.focus_set()

def cb():
    print e.get()

b = Button(m, text="get", width=10, command=cb)
b.pack()
mainloop()
```

19.2.5 `Frame` Widget

Frame is a rectangular graphical interface area showed on the screen. This widget is primarily used as a geometry master framework. It is also used to perform padding between other widgets. We can use the `frame` by invoking an object of `Frame` class. We can mention the `height`, `width`, `colormap`, `background`, and `relief` parameters for a `frame` object. Frame widgets can also be used for video overlay `place` holder or any other external process.

```
m = Tk()
Label(text="one").pack()
myfrm = Frame(width=760, height=530, bg="", colormap="new")
myfrm.pack(fill=X, padx=10, pady=10)
Label(text="two").pack()
mainloop()
```

19.2.6 `LabelFrame` Widget

The `LabelFrame` widget is another variant of the `Frame` that has been introduced in Tkinter 8.4 version. The basic phenomena of this `LabelFrame` are that it will always draw a borderline around the child widget. Such widget is mainly used when we want to group some similar types of the widget in a single group. To display the widgets in a group, first create a `LabelFrame` and add the widgets in that `LabelFrame`. The `frame` by default draws the border around the widget (Figure 19.6).

```
from Tkinter import *
m = Tk()
g = LabelFrame(m, text="Insert Text", padx=7, pady=7)
g.pack(padx=8, pady=8)
w = Entry(g)
w.pack()
mainloop()
```

Figure 19.6 shows the `label frame` widget.

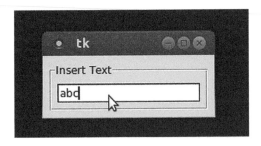

FIGURE 19.6
LabelFrame.

19.2.7 `Listbox` and `Menu` Widgets

19.2.7.1 *Listbox*

Listbox is a standard Tkinter widget that consists of text values in a list form. The listbox is generally populated with a list of text having the same font and font color. The user can choose one or more items from the listbox depending on the user's choice. Its primary use is to select a group of textual items at a time. During the time of construction of the listbox, it is by default empty, and hence, we have to insert the text on it. We can insert the text by the insert() method. The item that has been inserted is indexed from 0. Sometimes, we use special indexes such as ACTIVE that refers to the active items and END that appends the text at the end.

```
from Tkinter import *
m = Tk()
lb = Listbox(m)
lb.pack()
lb.insert(END, "Start list entry")
for i in ["1st value", "2nd value", "3rd value", "4th value"]:
    lb.insert(END, i)
mainloop()
```

Deletion of the list item is also possible. In most of the case, when we update the list, we use the delete() method for the listbox object.

```
lb.delete(0,END)
lb.delete(END,nw)
```

The listbox offers numerous selection modes. The major selection modes are SINGLE for selecting a single item from the list of items, BROWSE for selecting one or more items from a selection list through a mouse click, MULTIPLE for choosing multiple items by a single click, and EXTENDED that can be used by pressing Ctrl+Shift. The function listbx = Listbox(selectmode=EXTENDED) will perform an extended list selection process. The get() method is also used to get the list items for a given list-box. The get() method accepts both integer and string type, so conversion, in that case, is not required.

19.2.7.2 *Menu Widgets*

Menu widget supports pop-up, top-down, and top-level menu designs [4]. This makes all kinds of menus supported by the system. While implementing menu, we should take care of the design structure of the menu so that it will not create a fake menu. The following program is an implementation of the basic top-down menu bar that shows two menus at the top. No menu item is mentioned in this menu bar (Figure 19.7).

FIGURE 19.7
Pull-downmenu.

```
from Tkinter import *
r = Tk()
def first():
    print "This is First menu!!"
menubar = Menu(r)
menubar.add_command(label="Print", command=first)
menubar.add_command(label="Exit", command=r.quit)
r.config(menu=menubar)
mainloop()
```

In this case, as we click the Print menu, it will print the `message` in console and the Exit menu will quit the current window. A pull-down menu can be created using a similar technique. It will always be attached to the main menu. We can do this using the `add_cascade` operation. The following code describes the pull-down menu and its behavior:

```
from Tkinter import *
r = Tk()
def open():
    print "Opening mode"
def save():
    print "save document"
def cut():
    print "cut document"
def copy():
    print "copy document"
def paste():
    print "Paste Document"
def about():
    print "This is menu test"

menub = Menu(r)
f_menu = Menu(menub, tearoff=0)
f_menu.add_command(label="Open", command=open)
f_menu.add_command(label="Save", command=save)
```

```
f_menu.add_separator()
f_menu.add_command(label="Exit", command=r.quit)
menub.add_cascade(label="File", menu=f_menu)

# pull down menu creation
e_menu = Menu(menub, tearoff=0)
e_menu.add_command(label="Cut", command=cut)
e_menu.add_command(label="Copy", command=copy)
e_menu.add_command(label="Paste", command=paste)
menub.add_cascade(label="Edit", menu=e_menu)

h_menu = Menu(menub, tearoff=0)
h_menu.add_command(label="About", command=about)
menub.add_cascade(label="Help", menu=h_menu)

# display the whole menu in the window.
r.config(menu=menub)
mainloop()
```

A pop-up menu can be created in the same way as the pull-down menu. The only difference is that it can be displayed using the post () method.

```
menu.post(event.x_root, event.y_root)
```

Then, we can use the bind() method to canvas. The method bind() can do the same.

```
Frame.bind("button",popupmenu)
```

19.2.8 Radio Button

The radiobutton is a useful component in Windows programming. Tkinter supports radio buttons by means of radio button class. It is a kind of mutual exclusive button having many-to-one selection options. A radio button can contain either a text or an image. Any button can be associated with the function, and the call is made as the radio button choice is clicked. All the text in the button has a single font, and the control can be moved using TAB key. While setting Radio button, make sure that the button should be grouped and all buttons should point to the same variable (Figure 19.8).

FIGURE 19.8
Radio button.

```
from Tkinter import *
m = Tk()
MOD = [
        ("SVM", "1"),
        ("LR", "2"),
        ("KNN", "3"),
        ("CNN", "4"),
        ]
var = StringVar()
var.set("4")
for t, mod in MOD:
    b = Radiobutton(m, text=t,variable=var, value=mod)
    b.pack(anchor=W)
mainloop()
```

19.3 Concepts Covered in This Chapter

- Tkinter GUI components
- Button widgets
- Canvas widgets
- Listbox
- Menu
- Radio button

References

1. Lundh, Fredrik. "An Introduction to Tkinter." http://www.tcltk.co.kr/files/TclTk_Introduction_To_Tkiner.pdf (1999).
2. Shipman, John W. *Tkinter 8.4 Reference: A GUI for Python*. New Mexico Tech Computer Center, Socorro, NM, 2013.
3. Miller, Bradley N., and David L. Ranum. *Problem Solving with Algorithms and Data Structures Using Python*, Second Edition. Franklin, Beedle & Associates Inc., Portland, OR, 2011.
4. Grayson, John E. *Python and Tkinter Programming*. Vol. 140. Manning Publications, Greenwich, CT, 2000.

20

Python API Modules for Machine Learning and Arduino

Python library has a good range of API support for artificial intelligence and hardware interfacing. The machine learning application is one of the most popular packages that is widely implemented using python scikit learning tools. In this chapter, our main objective is to emphasize different useful packages that are related to machine learning, Arduino, Data Science, and other technologies. We are here trying to give the insights of those packages with some useful examples. One of the helpful entities in this context is PyPI, the Python Package Index, which is a software repository for Python programming.

20.1 Installation of the Package

Python package installer is called pip. In all Python versions, pip is available. Before using pip, first, we have to check the version of Python that is being currently used. This can be done using the following command:

```
python  --version
```

Then, we have to check the pip command. In the Linux environment, by default, this command is available, whereas in Windows, it may not be. Perhaps, to run this command in Windows, we need to go to the installation directory\Scripts subdirectory. Now we create a batch file local.bat to open a command line by writing cmd on the content of the batch. Now, we simply click that batch file, which will open the command window. From here, we can run the pip. We can check the version of pip by simply writing the pip version.

20.2 SciPy and NumPy Packages

NumPy is the basic package for scientific computing with Python [1,2]. This library provides Python some most powerful components such as multi-dimensional array objects. This includes all fundamental concepts of basic computations such as basic Fourier transform, linear algebra, statistical operation, sorting, selection, I/O, and other various logical and mathematical operations.

The core of the NumPy is the ndarray object. This is primarily used to encapsulate the *n*-dimensional homogeneous data. Some of the major differences between the NumPy array and the standard array in Python are as follows:

- Array in NumPyis of fixed data type. Therefore, one could not update the size of the array at all. Changing the size involves deleting the array and creating a new one.
- The data elements of NumPy array are of homogeneous type. Still, it supports an array of objects.
- Several complex scientific computations are possible in a NumPy array in a more efficient manner compared with a normal sequence-like list.

20.2.1 Creation of Shapes

The fundamental goal of NumPy is to create an array of *n* dimensions of homogeneous type. The dimension of the array is known as axis. The following example shows the operation of NumPy array in different ways:

```
>>> import numpy as np
>>> x=np.arange(20).reshape(4,5)
>>> x
array([[0,   1,   2,   3,   4],
       [5,   6,   7,   8,   9],
       [10, 11, 12, 13, 14],
       [15, 16, 17, 18, 19]])
>>> x.shape
(4, 5)
>>> x.ndim
2
>>> x.dtype.name
'int32'
>>> type(x)
<type 'numpy.ndarray'>
```

There are several types of techniques used to create an array. In NumPy, we use the array() function to create an array.

```
>>> import numpy as np
>>> n = np.array([1,3,5])
>>> n
array([1, 3, 5])
```

```
>>> a.dtype
dtype('int32')
>>> b = np.array([1.2, 3.5, 5.1])
>>> b.dtype
dtype('float64')
```

We can also explicitly specify the type of the array. The following piece of code shows the dimension supply explicitly for a two-dimensional array:

```
>>> z = np.array( [ [1,6], [3,7] ], dtype=complex )
>>> z
array([[1.+0.j, 6.+0.j],
       [3.+0.j, 7.+0.j]])
```

We can create a sequence of numbers using NumPy. The `arrange()` function returns an array of certain range of values, and the values are given as parameters.

```
>>> np.arange( 10, 30, 5 )
array([10, 15, 20, 25])
>>> np.arange( 0, 2, 0.3 )
array([ 0. ,  0.3,  0.6,  0.9,  1.2,  1.5,  1.8])
```

The first case shows an array of five elements between 10 and 30 with an interval of 5. The next case shows a floating point array of elements ranging from 0 to 2 with an interval of 0.3.

Sometimes, we need better predictable values within infinite floating point precision. To create numeric sequences, we use `linspace()` that receives the range and the number of values within that range. The following example illustrates the operation of `linspace()` function.*

```
>>> from numpy import pi
>>> np.linspace( 0, 3, 10 )              # 10 numbers from 0 to 3
array([0.    , 0.33333333, 0.66666667, 1.    , 1.33333333,
       1.66666667, 2.    , 2.33333333, 2.66666667, 3.    ])
>>> x = np.linspace( 0, 2*pi, 100 )   # useful to evaluate
function at lots of points
>>> f = np.sin(x)
```

20.2.2 Copy Operation in NumPy

The copy operation for an array might operate in two ways [3,4]. One the address of the list object gets copied into a target. Therefore, the same array objects are accessible from both the variables x and y.

---

* Code source: https://www.scipy.org/.

```
>>> x = np.arange (12)
>>> y = x                 # no new object is created
>>> y is x                # a and b are two names for the same
ndarray object
True
>>> y.shape = 3,4         # changes the shape of a
>>> x.shape
(3, 4)
```

A shallow copy of a list object can be made using view(). In this case, different objects can share the same data. But physically, memory will be created for separate objects.

```
>>> c = a.view ()
>>> c is a
False
>>> c.base is a
True
```

Finally, we use the copy() method to make an actual copy of one object. It will simply make two separate identical objects as the copy() method is invoked.

```
>>> y = x.copy ()              # a completely new object with new
data set has been created.
>>> y is x
False
>>> y.base is x               # y doesn't share anything with x
False
>>> y[0,0] = 9999
>>> x
array([[   0,   10,   10,    3],
       [1234,   10,   10,    7],
       [   8,   10,   10,   11]])
```

20.3 Introduction to Matplotlib

Matplotlib is the most powerful plotting library component in Python. Any type of data visualization, graph, or chart generation can be done using Matplotlib package. Designing of two- and three-dimensional graphs, bar charts, and pie charts is a very popular component of Matplotlib package. One of the advantages of Matplotlib is that it is a continuously evolving software, that is, new features and functionalities are ready to incorporate each time as it adds some new components. Some of the core features are as follows:

- It comprises 70,000 lines of code.
- It is a home to various interfaces and is able to interact with different back-end services.

20.3.1 Core Components

The core component of Matplotlib is Pyplot. It is a collection of functions that work with Matplotlib like any other professional plotting software. In Pyplot, numerous states are preserved across the function call such that it can keep track of various components such as figures and plotting areas. A simple line plot based on some predefined list data object is shown in Figure 20.1.

```
import matplotlib.pyplot as myplt
myplt.plot([2,4,6,8,7])
myplt.ylabel('some numbers')
myplt.show()
```

In this simple plot, the y value is the given data object and the x value is the number of objects in the list.

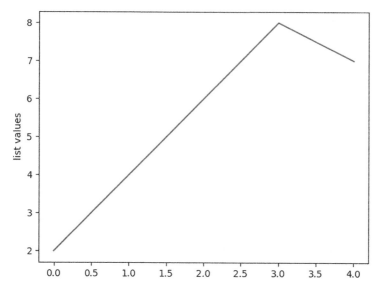

FIGURE 20.1
Plot function in Matplotlib.

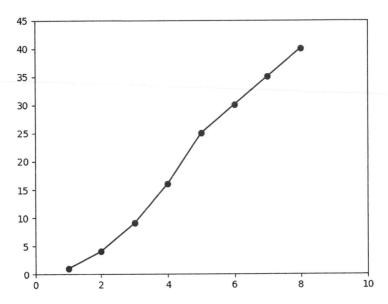

FIGURE 20.2
Colored dotted solid line graph.

plot() command is a very useful and versatile command. We can plot a graph of X vs. Y by simply supplying the list containing x and y coordinates. We can also pass the type of line in plotting through it. In this case, we provide the axis range by passing a list of values. plt.axis() takes the list of x and y axes. Here, 0–10 is the range for x-axis and 0–45 is the range for y-axis. "ro-" signifies the red colored, and "o" signifies the dotted solid line in Figure 20.2.

```
import matplotlib.pyplot as plt
plt.plot([1,2,3,4,5,6,7,8], [1,4,9,16,25,30,35,40], 'go-')
plt.axis([0, 10, 0, 45])
plt.show()
```

20.3.2 Line Properties

We can produce different types of lines in Matplotlib. This can be done by setting up several parameters of lines such as width, dash style, and antialiased effect. To set the width of the line, we can pass the line width parameter in the plot() method.

```
Plt.plot(a , b , linewidth=1.5)
```

Sometimes, we use the setter() method to set the properties such as the antialiased effect that deals with jaggies. Antialiasing has generally diminished the jaggies of the lines that should be smooth. These jaggies generally occur due to the improper resolution of the different output devices such as printer and monitor.

```
line, = plt.plot(x, y, '-')
line.set_antialiased(False)
```

The first method produces line objects with the corresponding x and y values with a–feature. In the second line, we call set_antialiased(False) to turn off antialiasing.

20.3.3 Multiple Plot Representation

Like MATLAB®, pyplot also supports multiple plots in a single figure. It comprises current figure and current axes. To create multiple plots in a single figure, we use the subplot() command. The figure() command can optionally be used to create a figure, as the figure will be taken by default. The subplot() command uses a number of rows and columns, with the figure number being a parameter. The figure number ranges from 1 to a number of rows × number of columns. Commas in subplot() are optional. So, subplot(211) is equivalent to subplot(2,1,1). A sample code for subplotting is shown as follows:

```
import matplotlib.pyplot as plt
plt.figure(1) # first figure explicit declaration
plt.subplot(211) # the first subplot in the first figure
plt.plot([8, 12, 13])
plt.subplot(212) # the second subplot in the first figure
plt.plot([14, 15, 16])
```

20.4 Overview of SciPy

SciPy is a Python-based ecosystem of an open source software bundle for mathematics, science, and engineering. It contains modules for optimization, linear algebra, integration, interpolation, signal and image processing, and fast Fourier transform.

To install SciPy, we use the pip install scipy command. The repository will automatically be downloaded by the system. The newest API release for SciPy is SciPy 1.1.0.

SciPy is a powerful tool for mathematical computation and analysis. SciPycan be used as an interactive tool for data processing and system prototyping tool along with MATLAB, R-Lab, SciLab, and many more. SciPy package comprises various subpackages as mentioned in Table 20.1.

TABLE 20.1

SciPy Package List

Cluster	Support for clustering algorithms
Constants	Support for physical and mathematical constants
fftpack	Fast Fourier transform routine and API support
integrate	Integration and ordinary differential equation solving API support
interpolate	Support for interpolation and smoothing splines
IO	Input and output routines
linalg	API for linear algebra
ndimage	API support for n-dimensional image processing
odr	Orthogonal distance regression support
optimize	Optimization and root-finding routines
Signal	Signal processing APIs and routines
sparse	Sparse matrices and their associated routines
spatial	Spatial data structures and their related algorithms
special	Special functions implementation API
stats	Statistical distributions and functions

20.5 Machine Learning Tools

The Python environment has a well-defined machine learning [5–7] tool package set to perform all complex machine learning tasks [8]. Numerous libraries are available freely to create a good and efficient machine learning code. To perform a machine learning operation, we have to follow some well-defined steps. The steps are as follows:

- Identify the problem definition.
- Prepare data set based on it.
- Apply some algorithms in those data sets.
- Get and optimize results.
- Represent the results.

A fundamental objective of machine learning technique is to load and summarize the data sets, evaluate algorithms, and make some predictions [9,10]. Therefore, to implement machine learning, the following mandatory packages need to be installed:

- NumPy
- SciPy

- Matplotlib
- sklearn
- pandas

We generally use `pip install` command to install libraries that implement the machine learning code. The sklearn package is mainly consist of different implementation of supervised learning algorithms in the form of a package module.

Before implementing any machine learning code, we should load the libraries. Some of the libraries that need to be imported are shown as follows:

```
import pandas
from pandas.plotting import scatter_matrix
import matplotlib.pyplot as plt
from sklearn import model_selection
from sklearn.metrics import classification_report
from sklearn.metrics import confusion_matrix
from sklearn.metrics import accuracy_score
from sklearn.linear_model import LogisticRegression
from sklearn.tree import DecisionTreeClassifier
from sklearn.svm import SVC
from sklearn.neighbors import KNeighborsClassifier
from sklearn.discriminant_analysis import
LinearDiscriminantAnalysis
from sklearn.naive_bayes import GaussianNB
```

Here, we can observe that we have imported different modules from pandas. `import matplotlib.pyplot` is used for the visualization of the graph. The sklearn package is mostly used to implement different classifiers. Typically, we use logistic regression, decision tree, support vector machine, K-nearest neighbor, and Gaussian and linear discriminant analyses.

We also generate confusion matrices and accuracy score values using `sklearn.matrices. confusion_matrix` and `accuracy_score`.

After doing this, we load the dataset. In this case, we typically use a standard iris dataset available in the UCI machine learning repository. We design the code for iris data set that is inspired by Jason Brownlee.

```
n_ame = ['sepal-length', 'sepal-width', 'petal-length',
'petal-width', 'class']
data_s = pandas.read_csv(test.csv, names=names)
```

As we load the dataset, we can get an idea of the instance of the data set often called as dimension. We can get the dataset instances of certain rows in the following way (50 instances in this case):

```
print(data_s.head(50))
```

Statistical description and class-based summary can also be done using the following methods:

```
print(data_s.describe())
print(data_s.groupby('class').size())
```

After getting some description of the dataset, we can do some plotting of data to show the characteristics of data. We can do bar, scatter, or line plot of the data set easily using Matplotlib library. The hist() function createsa histogram-type bar graph in this case.

```
data_s.hist()
plt.show()
```

After getting some dataset representation, we apply some machine learning algorithms to the dataset. To do so, we have to create a training and testing data set. The training data set is used to train the model that we are going to create, and the test data set is used to create an input to the model to validate the performance of the model.

There are several steps involved in it. First, we have to separate out the validation data set or test the dataset. We apply a k-fold cross-validation technique to it. The system k-fold cross–validation technique says that we have to make segmentation of the whole data set into k number of subsets and choose one among them as the test set with the remaining k−1 number of subsets joined together. In the next iteration, a different subset becomesthe test set, and the remaining k−1 will be the training set and so on. This improves the accuracy of the algorithm. In this case, typically we choose a tenfold cross-validation mechanism. We can build different models using different algorithms and finally select the best model.

Then, we have to go for the validation stage. In this case, we have to judge the model and review the performance of the model over some unseen data sets. Later, we use some statistical models to judge the accuracy of the results on some specific the model over some unseen data sets. The technique here is to split the dataset into two halves. 80% of the data is used for training, whereas 20% is used for validation.

```
array = dataset.values
X = array[:,0:4]
Y = array[:,4]
valid_size = 0.20
seed_1 = 7
X_train, X_validation, Y_train, Y_validation = model_
selection.train_test_split(X, Y, test_size=valid_size,
random_state=seed_1)
seed_1 = 7
score = 'accuracy'
```

In general practice, seed_1=7 produces the best result for all the algorithms; therefore, we can choose seed as 7.

Then, we have to apply different algorithms over these data sets. In this case, it is unknown that which technique gives an optimal result. For partially linearly separable data, it is expected that it produces a good result.

We now consider six models and apply them o the same dataset and get the result, that shows which model gives the best performance over the following data set:

```
my_models = []
my_models.append(('LR', LogisticRegression()))
my_models.append(('LDA', LinearDiscriminantAnalysis()))
my_models.append(('KNN', KNeighborsClassifier()))
my_models.append(('CART', DecisionTreeClassifier()))
my_models.append(('NB', GaussianNB()))
my_models.append(('SVM', SVC()))
res = []
n_ame = []
for n_ame, m_model in my_models:
        kfold = model_selection.KFold(n_splits=10,
random_state=seed_1)
        cv_results = model_selection.cross_val_score(m_model,
X_train, Y_train, cv=kfold, scoring=score)
        results.append(cv_results)
        n_ame.append(n_ame)
        my_msg = "%s: %f (%f)" % (name, cv_results.mean(),
cv_results.std())
        print(my_msg)
```

We can also do the plotting of the algorithms and show their performance in a candlestick plot. The algorithms are measured with the best accuracy using the tenfold cross-validation methodology (Figure 20.3).

```
fg = plt.figure()
fg.suptitle(' Comparison between algorithms')
axf = fg.add_subplot(111)
plt.boxplot(res)
axf.set_xticklabels(n_ame)
plt.show()
```

20.6 Arduino API for Python

Python serial interface is pretty straightforward for Arduino support. Like other serial interface devices, Arduino can be accessed through the serial

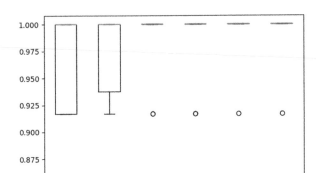

FIGURE 20.3
Algorithm comparison.

port and the baud rate. The following statement is enough for performing communication through serial ports.

```
>>> import serial
>>> s = serial.Serial('/dev/ttyUSB0', 9600)
>>> while True:
        print s.readline()
```

To add a data in the serial port of Arduino is very easy. We can invoke thes. write('a') method.

20.6.1 Arduino Prototyping API

Another useful API is known as Python Arduino Prototyping API. This API supports a rapid prototyping of Arduino without loading the program on the board repeatedly. To set up this API, first of all, prototype.pde should be loaded into the Arduino development board. Arduino lib should be imported to the Python script. The following code shows a basic blinking program written in Python and executed in Arduino:

```
#import the library first
from arduino import Arduino
import time
#specify the serial port on which Arduino is connected
my_board = Arduino('/dev/ttyUSB0')

#declare output pins as a list/tuple
my_board.output([11,12,13])
```

```
#perform operations
i=0
while(i<10):
    my_board.setHigh(13)
    time.sleep(1)
    my_board.setLow(13)
    time.sleep(1)
i+=1
```

20.6.2 Nanpy API

Nanpy API is mainly used to make a master–slave architecture between Arduino and the device that is interfaced with Arduino, such as PC or other microcontroller board. It comprises a bulk library support for the development of the prototype rapidly. Nanpy uses an expendable architecture through which it supports a wide range of libraries, such as a stepper, servo, LCD, OneWire, and Dallas temperature.

A simple code that demonstrates the use of nanpy is shown as follows:

```
from nanpy import ArduinoApi
x = ArduinoApi()
x.pinMode(11, x.OUTPUT)
x.digitalWrite(11, x.HIGH)
```

Interfacing an LCD screen in Nanpy is pretty easy. We simply mention the pin numbers and screen resolution as lists and pass them as parameters of the lcd() method.

```
from nanpy import Lcd
mylcd = Lcd([7, 8, 9, 10, 11, 12], [16, 2])
mylcd.printString('Hello Nanpy!')
```

The mechanism for detection of the serial port is also pretty simple. Nanpy has the feasibility to auto detect the serial port. However, we can also explicitly mention the serial port.

```
from nanpy import SerialManager
con = SerialManager(device='/dev/ttyACM1')
```

20.6.3 Nanpy Installation

Building and installing the firmware of Nanpy, we can do the following operation. The command is given as follows:

```
git clone https://github.com/nanpy/nanpy-firmware.git
cd nanpy-firmware
./configure.sh
```

To configure nanpy firmware, we can edit nanpy/cfg.h. To install Nanpy firmware, simply copy nanpy folder in the sketchbook folder and start Arduino IDE and upload sketchbook/nanpy. Then, click Upload.

In the master machine, you can just install Nanpy using the following command:

```
pip install nanpy
```

20.7 Concepts Covered in This Chapter

- NumPy and SciPy package
- Use of sklearn and pandas
- A sample machine learning code
- Python API for Arduino
- Nanpy API

References

1. Kiusalaas, Jaan. *Numerical Methods in Engineering with Python 3.* Cambridge University Press, Cambridge, 2013.
2. McKinney, Wes. *Python for Data Analysis: Data Wrangling with Pandas, NumPy, and IPython.* O'Reilly Media, Inc., Sebastopol, CA, 2012.
3. Pedregosa, Fabian, Gaël Varoquaux, Alexandre Gramfort, Vincent Michel, Bertrand Thirion, Olivier Grisel, Mathieu Blondel, et al. "Scikit-learn: Machine learning in python." *Journal of Machine Learning Research* 12, no. Oct (2011): 2825–2830.
4. Walt, Stéfan van der, S. Chris Colbert, and Gael Varoquaux. "The NumPy array: A structure for efficient numerical computation." *Computing in Science & Engineering* 13: 22.
5. Zemmal, Nawel, Nabiha Azizi, Nilanjan Dey, and Mokhtar Sellami. "Adaptative S3VM semi supervised learning with features cooperation for breast cancer classification." *Journal of Medical Imaging and Health Informatics* 6, no. 4 (2016): 957–967.
6. Zemmal, Nawel, Nabiha Azizi, Nilanjan Dey, and Mokhtar Sellami. "Adaptive semi supervised support vector machine semi supervised learning with features cooperation for breast cancer classification." *Journal of Medical Imaging and Health Informatics* 6, no. 1 (2016): 53–62.
7. Nath, Siddhartha Sankar, Girish Mishra, Jajnyaseni Kar, Sayan Chakraborty, and Nilanjan Dey. "A survey of image classification methods and techniques." In *2014 International Conference on Control, Instrumentation, Communication and Computational Technologies (ICCICCT)*, Kanyakumari, India, pp. 554–557. IEEE, 2014.

8. Machine Learning. Retrieved from https://developers.google.com/machine-learning/.
9. Brownlee, Jason. *Machine Learning Mastery with Python*. Machine Learning Mastery Pty Ltd., Vermont, Australia, 2016: 100–120.
10. Brownlee, Jason. "Machine Learning Mastery with Python: Understand Your Data, Create Accurate Models and Work Projects End-To-End." https://machinelearningmastery.com/products/ (2016).

21

More on Machine Learning API

Python scikit-learn gives a great utility to implement machine learning based intelligent computing. Along with that now a day there are wide verity of deep learning algorithms has also been introduced. In this chapter, our primary focus is on the different machine learning models that Python do support such as supervised learning and unsupervised learning including some well-known classes of algorithms. This chapter also includes the utilities of deep learning tools like Tensor Flow, Convolutional Neural Network (CNN) and Recurrent Neural Networks (RNN).

21.1 scikit-learn

Before we get deeper into the scikit-learn package, we have to understand the basics of machine learning.

The fundamental property of machine learning involves consideration of "n" samples of the data mainly used for training data. As a result, the machine learning algorithm can forecast the property of unknown data. Generally, the learning property can be categorized into specific classes, as discussed in Section 21.1.1.

21.1.1 Supervised Learning

The majority of machine learning algorithms use supervised learning mechanism [1,2]. In this case, we create an input variable x and an output variable y. We use a mapping function $y = f(x)$ from the input to the output as a learning function. The fundamental objective of supervised learning is to create a mapping function so that, for a new set of input data (x), one can forecast the output variable y. The technique is known as supervised learning because it depends on a training data set that is further used to train the function, and it is quite similar to perform the role of the teacher so that the function learns the features; if some test data is received, it can predict whether it is in the same or different class. In today's world, there are two types of machine learning techniques of prime interest:

- Classification: It is a technique that encompasses a feature of labeling some unlabeled data into some specified class. The data sample generally belongs to one or more classes. The function has to learn from the labeled data and then it must take a decision that the test data belongs to which class. A gender detection using face recognition is a classical binary classification problem, where there are two different labeled data. If the number of labels is more, then that is perhaps an n-ary classification. We can think that classification is a discrete form of supervised learning. Where one has a limited number of categories, for each category, there are k numbers of samples to be provided. The function should label them into the correct category of classes.

- Regression: It is a technique to deal with real values. In this case, the desired output is having one or more than one continuous variable. An example is to predict the age of animals from their height and weight.

Another form of learning is the unsupervised learning, where the training data set comprises the input vector x but no target value y defined. Clustering is one of the examples of unsupervised learning. Sometimes, the objective is to determine the data under a given input dimension. In such cases, data having higher dimension get synthesized into three-dimensional spaces, sometimes known as density estimation. In clustering, we focus on the grouping of data points. An example of grouping of peoples is an e-commerce website. On the other hand, association discovers the rule that describes the large portion of data. Some of the good examples are K-means clustering and a priori algorithm for rule learning problem.

Semisupervised learning is another category of learning. This category stays between unsupervised and supervised learning. In this case, for a large set of data x, some of the elements are labeled and the rest are not labeled. Major categories of practical cases fall under this category. In practical cases, it is pretty difficult to label a large amount of data because it is time-consuming and costly, and requires domain experts. On the other hand, unlabeled data are less costly and are easily available. In case of supervised learning, we use unlabeled data as training data and fit them into function and train the model. Then, we generate a prediction of unseen data from that model.

The scikit-learn package comprises a standard iris data set as a test set. We therefore can create and display the data set using the following Python statement:

```
from sklearn import datasets
i = datasets.load_iris()
d = datasets.load_digits()
print(i.data)
print(d.data)
```

The dataset is a dictionary structure. The whole dataset stores the data and some metadata about the data. Dataset stores the data in the form of a sample, feature array into a `.data` member. As shown in the previous code, `i.data` has the data members of the iris data set and `d.data` has the data members of the digit data set.

Learning and prediction are the key objective of the supervised machine learning models. In scikit-learn, the estimator for the classification is basically a Python object which implements two major methods `fit(predictor, target_value)` and `predict(test_dataset)`. The `fit()` method is primarily used for fitting the estimator. The estimator is a rule of computing a certain quantity based on a single observed data. The `predict()` method performs the prediction job based on the train data and the unknown test data.

In scikit-learn, a good example of the estimator is `sklearn.svm.SVC` that implements the support vector classification.

21.1.2 Support Vector Machine Classifier

Support vector machine (SVM) is the most popular supervised learning algorithm that is applicable to classification and regression tasks. In this methodology, we perform plotting of data points into an n-dimensional space. Then, our task is to find out a hyperplane that separates different classes in an efficient way (Figure 21.1).

The working principle of SVM is fundamentally based on identifying the hyperplane. Figure 21.2 shows the hyperplane selection mechanism.

In Figure 21.2a, there are three hyperplanes. We need to decide which plane we should choose. Here, the thumb rule is to select the hyperplane that better separates two different classes. In Figure 21.2a, plane B better segregates the two classes. In case of Figure 21.2b, hyperplane B is most suitable,

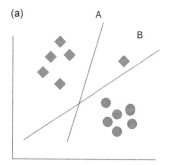
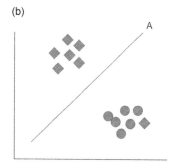

FIGURE 21.1
Classification case 2: (a) Classification type 1 and (b) classification type 2.

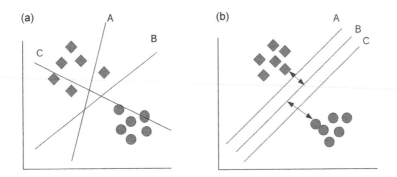

FIGURE 21.2
Classification case 1: (a) Hyperplane of different direction and (b) parallel hyperplane.

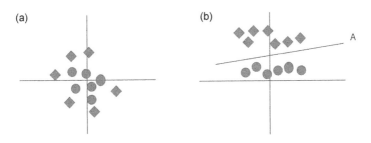

FIGURE 21.3
Classification case 3: (a) x,y plane; (b) x,z plane.

because plane B has the maximum distance from the nearest data point of two classes.

In the second case (Figure 21.3a), we have seen that one data point of a group is not close enough with the group of the same data point; therefore, two different planes can be thought of. Hyperplane A has a higher margin in comparison with that of B. However, we should choose plane B because hyperplane B has made better accurate classification in comparison with hyperplane A. In case of Figure 21.3b, we cannot segregate the two different data points of different classes using a straight line. In this case, SVM ignores the outliers.

In the next case, there is no straight line hyperplane existing for two classes. Although we have created the `linear` hyperplane for all cases, in this special case, the hyperplane is not `linear`. To solve this problem, a new feature of SVM has to be implemented that supports an additional plane z, where $z = x^2 + y^2$. Therefore, the plot of the data is now along the z and x axes. Figure 21.4a and b shows the classification of two different classes in the z, x hyperplane. In the plot, z is always positive because it is the square sum of x and y. SVM has a mechanism called kernel trick, which adds the feature automatically to have a new `linear` hyperplane.

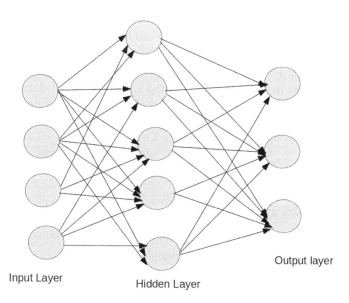

Input Layer

Hidden Layer

Output layer

FIGURE 21.4
The three-layer ANN visualization.

21.1.3 Implementation of SVM in Python

Python has an inbuilt SVM classifier called SVC in the scikit-learn package. The steps involved to create a prediction using SVM are importing the proper library, creating objects, fitting the useful model, and finally predicting the values.

```
Form sklearn import svm
my_model =  svm.svc( kernel = 'linear' , c= 1000, gamma =0.02)
my_model.fit(A,b)  # where A is predictor and b is target  for
training data set and
                              #a_test  is the predictor of
the test data set.
my_model.score(A,b)
result =  my_model.predict(a_test)
```

The parameter of the svc model has a significant meaning. Different parameters of the SVM classifier are discussed as follows:

- kernel: It generally acts as a similarity function. It takes two inputs and splits out how similar they are. Different types of kernel functions, such as linear, radial basis, and polynomial, are highly useful for linear and nonlinear hyperplanes. The default kernel function is the radial basis function.

- c-value: It is a penalty parameter of the error term. This value is responsible for controlling the trade-off between the correct classification of training points and the decision boundary.
- gamma: This is the kernel coefficient of radial basis, linear, and sigmoid functions. If the gamma is high, the test data set tries to exactly fit with the train data set to avoid the overfitting problem.

Some of the other important parameters are cache_size, class_weight, coef0, decision_function_shape, degree, max_iter, probability, random_state, shrinking, tol, and verbose.

21.1.4 Advantages and Disadvantages of SVM

- SVM works in a clear margin of separation of data points.
- It is highly useful for higher dimensional space.
- In the decision function, it only uses a subset of training points, and hence, it is highly efficient in terms of memory.
- When the number of dimensions is greater than the number of samples, it works really better.
- For a large dataset, the performance of SVC is compromised because of its high training time.
- For noisy data set where the target class overlaps, the model suffers from a serious issue.
- Probability estimation could not be obtained by SVC directly. It can be obtained from k-fold cross-validation.

21.1.5 Artificial Neural Networks

Artificial neural network (ANN) is a very important component of machine learning [3,4]. It primarily mimics the operation of human neurons. The main component of the ANN involves the input, hidden, and output layers. A standard diagram of ANN is shown in Figure 21.5. Just like atoms that are fundamental to the neural network (NN), perception is the backbone of all NNs.

Perceptron can be thought of anything that can take the input values of multiple numbers and give an output of a single value. The structure shown in Figure 21.6 takes three inputs and hence produces one output.

The defining relationship between the input outputs in the perceptron is an important point. One way is the direct combination of the inputs and summing up them and setting a threshold equal to 0. If the threshold is crossed, then the output is 1; otherwise, it is 0. For example, suppose

$$a = 1, b = 1, c = 0$$

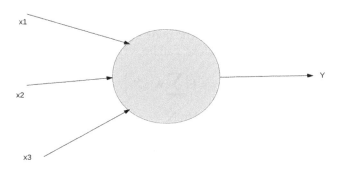

FIGURE 21.5
Three-input perceptron.

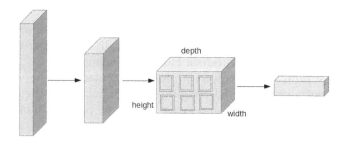

FIGURE 21.6
CNN arrangement in a three-dimensional space.

Therefore, a+b+c > 0 should be 1 that crosses the threshold so that the output value must be 1 in this case.

In the next case, we add weight to the inputs to compute the outputs. For example, we can put the weights $w1 = 1$, $w2 = 2$, $w3 = 3$ for the coefficients x,y,z so that $w1 \times x + w2 \times y + w3 \times z >$ threshold. To compute the output, we must multiply the weight values with each input. And finally, it is judged whether the value crosses the threshold.

In the next case, we add a bias value along with the summation of the weighted input values. The biasness of the perceptron reflects how much flexible the perceptron is. It is somehow similar to the function $y = mx+c$. Therefore, we can just add a bias value to the equation $w1 \times x + w2 \times y + w3 \times z + 1 \times b >$ threshold.

21.1.6 Activation Function

Till now, the perceptrons are only dealt with using the `linear` function. The nonlinear transformation of the perceptron is, however, a major issue. A neuron that uses nonlinear transformation basically uses the activation function.

The activation function basically takes the sum of weighted inputs as parameters and therefore returns the neuron itself.

The activation function, therefore, can be written as follows:

$$y = F\left(\sum w_i \cdot X_i\right)$$

$$i = 0 \text{ to } N$$

The main goal of the activation function is to perform a nonlinear transformation to fit a nonlinear hypothesis or to estimate a nonlinear complex function. There are various activation functions used in this scenario, some of which are ReLu and sigmoid.

21.1.7 ANN Implementation Using Python API

We can develop an ANN model using Python API [5]. One of the most popular APIs in this context is Keras [6,7]. It is basically a wrapper of the most efficient NN libraries such as Tensor Flow and Theano.

There are several steps involved in the implementation of an NN: loading the dataset, defining the model, compiling the model, fitting the model, and finally evaluating the model and tying the whole process altogether. The core requirement to run the ANN is to install numpy, scipy, and keras modules. All the modules should be configured in a proper way.

It is a thumb rule that runs a machine learning algorithm to start with a stochastic process which is nothing but a random seed. This technique is effective to run the same code repeatedly, and hence, the same result will be obtained, which is useful for analysis and comparison of the results having similar randomness.

```
from keras.models import Seq
from keras.layers import Den
import numpy as np
np.random.seed(7)
```

Our second task is to load a data set and split it into two different parts, namely, input variable X and output variable Y. The data set may be any data having a .csv format or XML.

```
dataset = np.loadtxt("data_set.csv", delimiter=",")
X = dataset[:,0:8]
Y = dataset[:,8]
```

Then, we have to define the model. The model in Keras comprises several layers. We can add layers until the topology is satisfactory. We have to ensure that the input layer has the right number of inputs. To do so, we have to use the input_dim argument.

```
m = Sequential()
m.add(Dense(10, input_dim=6, activation='relu'))
m.add(Dense(8, activation='relu'))
m.add(Dense(1, activation='sigmoid'))
```

Here, the Dense class is used to refer the fully connected network. We have to specify the number of neurons in the first argument. input_dim parameter sets the first layer with six input variables. We have to choose a specific activation function using the activation parameter. The rectifier "relu" activation function is applied to the input and hidden layers, whereas the sigmoid function is applied to the output layer. Rectifier activation performs better than sigmoid, and the output layer is a sigmoid because we want to map the output value in the map of 0 and 1.

Now, we must compile the defined model. Compilation involves the inclusion of libraries such as Tensor Flow. In the background, it will automatically choose the best model and run the training using CPU or graphics processing unit (GPU). During training, the loss function must be specified to evaluate a set of weights. The optimizer automatically searches through the weight of the network.

```
m.compile(loss='binary_crossentropy', optimizer='adam',
metrics=['accuracy'])
```

The optimizer that is used here is a gradient descendant optimizer. The third argument is metrics, which is nothing but accuracy. Then, we can train or fit the loaded data by calling the fit() function. The fitting process has to be run by a fixed number of iterations called epoch. We have to mention that using the epochs argument. We can also mention the number of instances that can be evaluated using batch_size.

```
m.fit(X, Y, epochs=150, batch_size=10)
```

After training, it's time to see the performance of the model for the dataset given. One can compute the performance of the model using the evaluate() method.

```
s = m.evaluate(X, Y)
print("\n%s: %.2f%%" % (m.metrics_names[1], s[1]*100))
```

21.2 Deep Learning Tools

Deep learning is a special form of machine learning in which the machine learns the natural behavior as a human [8–11], for example, a rocket autopilot unit and a driverless car. It is also a key component of voice control devices such as voice control elevator and TV. The deep learning system is fundamentally based on the deep NN. A very simple deep learning example we can say is the predictive analysis.

The system that uses the deep learning mechanism basically in each hierarchy applies some nonlinear transformation over its input and uses what it learns and tries to build a statistical output from it. The iteration continues until a satisfactory amount of accuracy in the output level is reached.

In traditional machine learning, the most challenging task is the identification of the proper feature to properly learn the computer. This task is called the feature extraction, and the success of the algorithm highly depends on the accurate definition of the feature set by the developer. In case of deep learning, the program itself builds the feature set without any supervision; therefore, it is quite similar to unsupervised learning. Initially, the system has given some initial labeled data. The program takes those basic data and tries to create an initial feature set and build a predictive model. But in that case, any similar object having almost a similar feature may be labeled as the same element. A system that deals with deep learning is sometimes called a deep NN, and it will take weeks or months to train the computer with thousands to millions of training data (mainly image). To achieve a good level of accuracy, deep learning needs an immense amount of data and a huge processing power. Nowadays, by virtue of cloud computing, IoT, and big data, it is quite possible to achieve a good level of processing power and therefore a good use of deep learning technology.

21.2.1 Convolutional Neural Network and Recurrent Neural Network

21.2.1.1 Convolutional Neural Network

As discussed in the previous section (section 21.2), convolutional neural network (CNN; [12]) is similar to a traditional ANN with weights and biases [13–16]. Each perceptron performs some nonlinear transformation of the input data by performing a dot product. The network is expressed by a single score function having input as raw image pixels and output as score value.

A normal NN just receives regular inputs and passes them through the hidden layer that consists of neurons that are fully connected to the previous layer. At the output layer, the classification value is represented by the classification score. Regular ANN never scales the full images, but one of the advantages of CNN is that it exploits the three-dimensional model of NN. The architecture is such that the image it takes as an input for processing in more sensible way. Unlike normal ANN, CNN systems take the image data in the form of height, width, and depth.

For example, the input images in CIFAR-10 are an input volume of activations, and the volume of the activations has dimensions $32 \times 32 \times 3$ (width, height, and depth, respectively). The neurons in a layer will only be connected to a small region of its previous layer, instead of all of the neurons in a fully

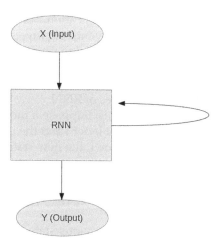

FIGURE 21.7
AnRNN.

connected manner. However, the final output layer for CIFAR-10 has dimensions $1 \times 1 \times 10$, and the reason is that by the end of the ConvNet (CNN) architecture, the full image is reduced into a single vector of class scores, arranged along the depth dimension. The visualization is shown in Figure 21.6.

A simple ConvNet is a series of layers. Every layer of CNN transforms one activation volume to another using differentiable functions. Three main layers are involved in building the ConvNet: convolution layer, pooling layer, and fully connected layer. An example sequence for the processing of CIFAR-10, the CNN should build through the following layered architecture:

- INPUT: This layer holds the input pixel values of the image ($32 \times 32 \times 3$) with three color channels (RGB).
- CONV: This layer computes the output of local neurons. Here, each neuron computes the dot product between their weights.
- RELU: This layer is an efficient element wise activation function. When it thresholds at 0, the original volume size will not change.
- POOL: This layer performs a downsampling operation.
- FC: This is a fully connected layer, mainly the outer layer, that will compute the scores of the class. Each neuronis fully connected with its previous layer.

21.2.1.2 *Recurrent Neural Network*

Recurrent neural network (RNN) is a type of NN that has diversified applications. It specifically deals with various input and output types. One of the important uses of RNN is sentiment classification. So, it is often used to classify the tweeter feed as having positive and negative sentiments. This NN model is also used for the language translation operation.

A simple example can be set such that suppose a word "hello" and the input are given, only the four letters h, e, l, l, and the network should predict the letter o. It is one of the practical uses of RNN.

The recurrent formula works like this: Suppose a model wants to predict the letter o, then it will take the current letter l and the previous letter which is also l; therefore, for an instant of time t, the new state can be obtained by following the recurrent formula:

$$H_{t+1} = f(H_t, X_{t+1})$$

Where H_{t+1} is the new state, H_t is the previous state, and X_{t+1} is the current input. Therefore, if we create the activation function for the same, it will be written as

$$H_{t+1} = \tanh(W_{hh}H_{t+1,} + W_{xh}X_{t+1})$$

Where W_{hh} is the weight of the previous state and W_{xh} is the weight of the current input. The Figure 21.7 depicts a RNN implementation.

21.2.2 Tensor Flow API for Deep Learning

Tensor Flow is an open source library for machine learning for the researcher as well as for the developers [17]. The API is organized in such a way that it supports beginner desktop application to web, cloud, and many more. The highest level API of Tensor Flow provides building blocks for creating and training deep learning models. The Tensor Flow model was created and maintained by Google. This has been run under Apache 2.0 open source license.

The Tensor Flow can run in a stand-alone CPU, GPU mobile device, and cloud. It is even designed in such a way that it can also run a large-scale cluster.

The Tensor Flow model has three components:

1. Nodes: They perform the computation over 0 or more than that number of inputs and outputs. Data must pass through the node in the form of multidimensional array and is identified as a tensor.
2. Edges: These are the connectors among various states. They provides the flow of data, branching, and looping, and some special edges do the synchronization of the state within the graph.
3. Operation: The operation named as an abstract computation takes an input value and the corresponding output value.

We can install the sensor flow package in Ubuntu by installing a virtual environment. To get this, we should use the following commands:

```
sudo apt-get install python-pip python-dev python-virtualenv
# for Python 2.7
sudo apt-get install python3-pip python3-dev python-virtualenv
# for Python 3.n
```

To get a higher version of pip, we use the following command:

```
sudo pip install -U pip
```

We have to create a directory to install the virtualenv using the following commands:

```
mkdir ~/tensorflow
cd ~/tensorflow
 virtualenv --system-site-packages venv            # for
python default (Python 2.7)
 virtualenv --system-site-packages -p python3 venv #  Python 3.n
```

Next, we should make the virtual environment active as follows:

```
source ~/tensorflow/venv/bin/activate   # bash, sh, ksh, or zsh
 source ~/tensorflow/venv/bin/activate.csh  # csh or tcsh
 . ~/tensorflow/venv/bin/activate.fish      # fish
```

Finally, we install Tensor Flow under the virtual environment using the following command:

```
pip install -U tensorflow
```

A basic Tensor Flow model can be implemented as follows by importing the tensorflow package:

```
import tensorflow as tf1
s1 = tf1.Session()
a = tf1.constant(10)
b = tf1.constant(32)
print(s1.run(a*b))
```

The detailed API reference is available in thetensorflow.org website.

A sample Keras-based NN model using Tensor Flow is shown in the following text. Here, the NN model comprises four layers. In this model, the second and fourth layers are fully connected. During compilation, the adam optimizer is taken, which is a stochastic gradient descendant optimizer giving the best performance for a multilayered NN. The loss function is `sparse_categorical_crossentropy` as the target is an integer. The code is shown as follows:

```
# courtesy tensorflow.org
import tensorflow as tf
mnist = tf.keras.datasets.mnist

(x_train, y_train),(x_test, y_test) = mnist.load_data()
x_train, x_test = x_train / 255.0, x_test / 255.0

model = tf.keras.models.Sequential([
  tf.keras.layers.Flatten(),
  tf.keras.layers.Dense(512, activation=tf.nn.relu),
  tf.keras.layers.Dropout(0.2),
  tf.keras.layers.Dense(10, activation=tf.nn.softmax)
])
model.compile(optimizer='adam',
              loss='sparse_categorical_crossentropy',
              metrics=['accuracy'])

model.fit(x_train, y_train, epochs=5)
model.evaluate(x_test, y_test)
```

21.2.3 The PyTorch API

The most advanced computation library in the machine learning domain is PyTorch. PyTorch is supposed to be an easy-to-use and easy-to-understand API ever created by a human being for machine learning.

The philosophy of API is that one can run it very easily and immediately. We do not have to wait for the whole code to be written and executed at a glance. The fundamental phenomena are described as follows:

- It is easy to implement and use.
- It can smoothly integrate with the data science stack of Python. It has enough similarity with NumPy.
- Dynamic computation platform is supported by PyTorch. PyTorch framework builds a computational graph as we start using it, and the graph can dynamically be changeable during runtime.

It also has a good advantage of multi-GPU support. As the framework was developed in 2016, data scientists may take leverage of the same. The use of PyTorch is best fitted for a replacement of NumPy using a powerful GPU. It also builds a deep learning research platform that provides maximum speed and flexibility. PyTorch provides a tensor that lives on either GPU or CPU or both. There are various tensor routines to accelerate the tensor routines that enhance scientific computation. There are faster executions of math, `linear` algebra, indexing, and slicing (Figure 21.8).

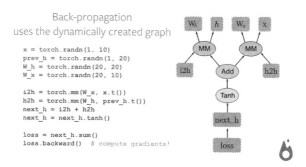

FIGURE 21.8
PyTorch execution environment. (Source: pytorch.org.)

A tape recorder approach is significantly used for creating and building a dynamic NN. One of the advantages of PyTorch is its reverse-mode automatic differentiation that allows the network to change its behavior with a minimum latency and overhead, which cannot be done in the case of Tensor Flow and Theano. PyTorch is deeply integrated into Python. It can be used naturally like NumPy, SciPy, and scikit-learn. PyTorch API is designed to be linear in thought and easy to use. For large NNs, PyTorch runs significantly fast. The system also increases the memory efficiency.

21.2.4 PyTorch Installation

To install PyTorch API in Windows, we can use open source installer conda. To get a conda installer, we have to download Anaconda in the system. We can download Anaconda distribution from the https://www.continuum.io/ downloads website. We have to download the right Python distribution and it is recommended to download and install the .exe file in Windows. Next, we have to download the Kivy installation wheel. Then, we have to anaconda in the command prompt and write the following commands to download the PyTorch API:

```
conda install -c anaconda python=3.6.1
conda install -c peterjc123 pytorch=0.1.12
```

After doing this, the Kivy installation should be run. Just change the path to kivy downloaded folder. Now, use the following command in Anaconda console:

```
pip install docutils pygments pypiwin32 kivy.deps.sdl2 kivy.
deps.glew
pip install kivy.deps.gstreamer
pip install Kivy-1.10.1.dev0-cp36-cp36m-win_amd64.whl
```

You can check the installation using the following command:

```
conda list
```

21.2.5 PyTorch Tensor Implementation

NumPy is an ultimate framework for machine learning, but it doesn't use the GPU to accelerate the high-end numerical computation [18]. In the modern era of deep learning computation for a huge chunk of data as well as image or video, a GPU provides a speedup up to 50× or greater than that.

As NumPy cannot exploit the feature, the most fundamental PyTorch concept is introduced in the form of tensor. The tensor is conceptually identical with the NumPy array. PyTorch offers numerous functions for operating such tensors. Tensor also keeps track of the gradient and the computational graph. These are perhaps the smartest tools for scientific computing. As PyTorch supports GPU, we have to simply cast a new data type that runs for tensor in a GPU-guided environment. A two-layer tensor using PyTorch is discussed in the following code example:

```
# code courtesy pytorch.org
import torch
dt = torch.float
dev = torch.device("cpu")
# device = torch.device("cuda:0") # Uncomment this to run on GPU

# N is batch size; D_in is input dimension;
# H is hidden dimension; D_out is output dimension.
N, D_in, H, D_out = 64, 1000, 100, 10

# Create random input and output data
x = torch.randn(N, D_in, device=dev, dtype=dt)
y = torch.randn(N, D_out, device=dev, dtype=dt)

# Randomly initialize weights
w1 = torch.randn(D_in, H, device=dev, dtype=dt)
w2 = torch.randn(H, D_out, device=dev, dtype=dt)

lrning_rate = 1e-6
for t in range(500):
    # Forward pass: compute predicted y
    h = x.mm(w1)
    h_relu = h.clamp(min=0)
    y_pred = h_relu.mm(w2)

    # Compute and print loss
    loss = (y_pred - y).pow(2).sum().item()
    print(t, loss)

    # Backprop to compute gradients of w1 and w2 with respect
to loss
```

```
grd_y_pred = 2.0 * (y_pred - y)
grd_w2 = h_relu.t().mm(grad_y_pred)
grd_h_relu = grad_y_pred.mm(w2.t())
grd_h = grad_h_relu.clone()
grd_h[h < 0] = 0
grd_w1 = x.t().mm(grd_h)

# Update weights using gradient descent
w1 = w1 - lrning_rate * grd_w1
w2 = w2 - lrning_rate * grd_w2
```

First, we have to select the device. By default, the code will run in the CPU. If we want to run it in the GPU, we simply give `torch.device()` parameter as a specific GPU. We have to provide batch size, input dimension, hidden dimension, and output dimension. Then, we have to provide random input and output data and initialize the weights. After weight initialization, compute loss and the predicted value of y. Finally, we apply back propagation to compute the gradients w1 and w2 with respect to loss, and update the weights.

21.3 Concepts Covered in This Chapter

- Supervised learning techniques
- SVM Python API
- NN Python API
- Deep learning tools CNN and RNN
- Python Tensor Flow API
- PyTorch implementation for deep learning

References

1. Dey, Nilanjan, Amira S. Ashour, and Surekha Borra, eds. *Classification in BioApps: Automation of Decision Making.* Vol. 26. Springer, Berlin, Germany, 2017.
2. Ketkar, Nikhil. "Introduction to Pytorch." In *Deep Learning with Python,* pp. 195–208. Apress, Berkeley, CA, 2017.
3. Chatterjee, Sankhadeep, Sirshendu Hore, Nilanjan Dey, Sayan Chakraborty, and Amira S. Ashour. "Dengue fever classification using gene expression data: a PSO based artificial neural network approach." In *Proceedings of the 5th International Conference on Frontiers in Intelligent Computing: Theory and applications,* pp. 331–341). Springer, Singapore, 2017.
4. Chatterjee, Sankhadeep, Sarbartha Sarkar, Nilanjan Dey, Amira S. Ashour, Soumya Sen, and Aboul Ella Hassanien. "Application of cuckoo search in

water quality prediction using artificial neural network." *International Journal of Computational Intelligence Studies* 6, no. 2–3 (2017): 229–244.

5. Lan, Kun, Dan-tong Wang, Simon Fong, Lian-sheng Liu, Kelvin K. L. Wong, and Nilanjan Dey. "A survey of data mining and deep learning in bioinformatics." *Journal of Medical Systems* 42, no. 8 (2018): 139.

6. Keras. Keras: The Python Deep Learning Library. Retrieved from https://keras.io/.

7. GitHub. Retrieved from https://github.com/keras-team/keras.

8. Chollet, Francois. *Deep Learning with Python*. Manning Publications, Greenwich, CT, 2017.

9. den Bakker, Indra. *Python Deep Learning Cookbook: Over 75 Practical Recipes on Neural Network Modeling, Reinforcement Learning, and Transfer Learning Using Python*. Packt Publishing, Birmingham, 2017.

10. Hu, Shimin, Mengyu Liu, Simon Fong, Wei Song, Nilanjan Dey, and Raymond Wong. "Forecasting China future MNP by deep learning." In *Behavior Engineering and Applications*, pp. 169–210. Springer, Cham, Switzerland, 2017.

11. Dey, Nilanjan, Simon Fong, Wei Song, and Kyungeun Cho. "Forecasting energy consumption from smart home sensor network by deep learning." In *International Conference on Smart Trends for Information Technology and Computer Communications*, pp. 255–265. Springer, Singapore, August 2017.

12. Yin, Wenpeng, Kann, Katharina, Yu, Mo, and Schutze, Hinrich. *Comparative Study of CNN and RNN for Natural Language Processing*. Retrieved from https://arxiv.org/abs/1702.01923, 2017.

13. Li, Zairan, Nilanjan Dey, Amira S. Ashour, Luying Cao, Yu Wang, Dan Wang, Pamela McCauley, Valentina E. Balas, Kai Shi, and Fuqian Shi. "Convolutional neural network based clustering and manifold learning method for diabetic plantar pressure imaging dataset." *Journal of Medical Imaging and Health Informatics* 7, no. 3 (2017): 639–652.

14. Li, Zairan, Nilanjan Dey, Amira S. Ashour, Luying Cao, Yu Wang, Dan Wang, … and Shi Fuqian. "Convolutional neural network based clustering and manifold learning method for diabetic plantar pressure imaging dataset." *Journal of Medical Imaging and Health Informatics* 7, no. 3 (2017): 639–652.

15. Wang, Yu, Yating Chen, Ningning Yang, Longfei Zheng, Nilanjan Dey, Amira S. Ashour, … and Shi Fuqian. "Classification of mice hepatic granuloma microscopic images based on a deep convolutional neural network." *Applied Soft Computing* 74 (2019): 40–50.

16. Hore, Sirshendu, Shouvik Chakraborty, Amira S. Ashour, Nilanjan Dey, Ahmed Ashour, Dimitra Sifaki-Pistolla, … and Sekhar Ranjan Bhadra Chaudhuri. "Finding contours of hippocampus brain cell using microscopic image analysis." *Journal of Advanced Microscopy Research* 10, no. 2 (2015): 93–103.

17. TensorFlow. Retrieved from https://www.tensorflow.org/.

18. Tang, Raphael, and Jimmy Lin. Honk: A PyTorch reimplementation of convolutional neural networks for keyword spotting. *arXiv preprint arXiv:1710.06554* (2017).

22

Conclusion

22.1 Illustrations of Chapters 1 and 2

Chapter 1 represents the concept of the open source hardware. In this chapter, primarily Arduino and its ecosystems are discussed. The different models of Arduino are been addressed in this context. Along with the hardware, the Arduino software suite is considered. The installation of the drivers and the IDE itself is mentioned. Chapter 2 involves a detailed illustration of the Arduino board architecture. The architecture of the AVR microcontroller, which is the core of the Arduino, is considered. Power units, digital output and input, and several allowable peripherals are also considered. Some of the major concepts such as interrupts, third-party device interfacing, and allowable libraries are also discussed in this chapter.

22.2 Illustrations of Chapters 3–5

Chapter 3 is a dedicated chapter for Arduino programming methodology. Different data types that are used in Arduino and the declaration of variables, operators, and their precedence are discussed. Functions and modular programming approaches are addressed in Chapter 4, for example, setup() and loop() functions. Recursive method call, library function design, and the fermata library concept are addressed in this context. Chapter 5 comprises conditional statements such as if-else, loop, and switch-case systems. Arduino is a software-driven embedded computation platform. The fundamental firmware program design is the main goal, which is successfully fulfilled in this chapter.

22.3 Illustrations of Chapters 6–9

Chapters 6 and 7 basically deal with the input and output methodology for Arduino with several examples. In Chapter 6, the pinMode() function, which is no doubt an important function, is discussed. Along with that standard input, functions are analyzed with their ease of use. In Chapter 7, the output mechanisms are addressed. Functions such as `digitalWrite()` and `analogWrite()` are discussed closely. An application of `Serial.println()` is also demonstrated. In this case, a Python interface is created to visualize the output of the `Serial.println()` function. Chapter 8 involves a unique open source visualization tool called Processing. The various aspects of a processing tool in graphical and animated visualization are addressed.

22.4 Illustrations of Chapters 10–21

Chapters 10–21 are specially designed for the beginners of the Python language. The concept of the Python language is discussed in Chapter 10, and the fundamentals of operators, variables, and expression in Chapter 11. Chapter 12 involves the decision-making and flow of control. Various decision statements such as if-else and loops such as while and for are emphasized in Chapter 12. Chapter 13 involves the functions and modular programming approaches used in Python. Several specialized approaches are covered in this context. Chapters 14 and 15 give a more realistic approach to implement a set of APIs and the well-known data structure. The primary focus of Chapter 14 is to use several APIs to access CSV and JSON data. Besides, an interaction with the MongoDB database system is also addressed in this case. Chapter 15 discusses the most classical types of data structures that are present in Python such as list, dictionary, and tuples. Chapter 16 gives the essence of the object-oriented paradigm of the Python language. How the properties such as inheritance and polymorphism have been addressed is discussed in this context. Further, Chapter 17 is all about the input–output scheme in Python, and Chapter 18 gives an essence of exception-handling mechanism in Python. Chapters 19–21 provide more advanced concepts such as fundamentals of GUI programming (Chapter 19). Chapters 20 and 21 give a glimpse of the machine learning APIs and some real-life examples. The introduction to the concept of deep learning methodology is also incorporated.

22.5 Future Scope & Enhancement of This Book

This book basically covers the introductory theories and the implementation of open source hardware and software paradigms. The students who use this book are skilled to design Arduino-based systems as well as Arduino-driven systems that have an interface of Python. Data acquisition using a sensor network and real-time data analysis is one of the major domains nowadays. This book highlights the potential implementation and design of such a system that may be used for the further tool in the field of real-time data analysis and predictions. Still, there are lot more enhancements done by introducing a detailed implementation of machine learning and deep learning approaches and tools.

22.5.1 The Final Wrap-Ups

In this guidebook, several modern intelligent application developments with open source environment have been addressed. From the beginning of this book, the introduction to Arduino and the open source prototype development has been discussed. Various software and hardware tools related to Arduino have been taken into account. Besides, Python—the most popular programming language—and its application development environment have also been considered. Finally, some of the most interesting artificial intelligence and machine learning tools have been emphasized. This book gives a direction towards the horizon of latest machine learning and deep learning techniques along with the open source prototyping platforms.

So, let's make some intelligent things.

Frequently Asked Questions

1. What if I have not had any prior experience in programming and hardware?

   The platforms such as Arduino and Python are easy enough to learn. They are designed in such a way that they support a self-learning platform. Both Python and Arduino community have strong resources and supporting documentation so that we can learn them in a pretty faster way.

2. Can Arduino run in any operating system (OS)?

   Arduino supports different types of OS platform such as Windows, Linux 32 and 64 bit, and MAC OSX Lion. Also, it is compatible with ARM-based Linux, which means it can also run in Raspberry Pi platform.

3. Is Python an object-oriented programming (OOP)?

   Python is said to be an object-oriented language from earlier days. It is said to be an object-oriented program (OOP) because of its support such as class, polymorphism, inheritance, and method overriding. It perhaps contains the core feature of OOP as well as procedure-oriented language.

4. Is there any need of an Arduino driver for Ubuntu Linux?

   In case of Linux distribution, the most recommended .tar.bz version of Arduino is widely used. In general case, as the Arduino IDE is opened and the hardware gets connected with the USB, the USB device gets automatically detected as /dev/ttyACM0 or /dev/ttyUSB0. As the Arduino is connected, it finds and lists out all possible USB ports connected with Arduino. If a match is found with product+vendor ID, then the board will appear with the serial port menu entry name.

5. What are the different sensors we use with Arduino?

   Arduino supports various analog as well as digital sensors, starting from a simple LDR to LM35 as well as temperature humidity and many more. The digital sensors such as motion sensor, ultrasonic sensor, and gas sensor are popular among them. They support MEMS devices such as gyroscope, accelerometer, magnetic compass, and barometer.

6. Why should we use Processing?

   Processing is a useful programming environment for visualization of real-time data. It supports a wide range of colors and effects, which are greatly useful not only for visualization but also for creating a unique visual effect. The fundamental language in the development environment is Java. The core application simplifies the fundamental multimedia support, such as OpenGL, PDF, and Camera capture. Processing uses extensible library supports and code structure to create a bunch of 3D complex geometries.

7. Up to what voltage Arduino won't burn (the operating voltage)?

The voltage range for Arduino is between 5 and 20 V. If less voltage is given to the Arduino, the activity of the board becomes very unstable. In such case, the board will not provide enough power to its peripheral devices such as input sensors or output actuators. The voltage over 12 V will affect the functionality of the power regulator inside Arduino. Within 18–19 V, the regulator starts burning and the Arduino board gets damaged.

8. When should we use for loops of Python?

It is best to use a for loop for iterating over a sequence of numbers. The for loop takes each element of the sequence and stores into an iteration variable. Another best use of the for loop in Python is to iterate over the list structure.

9. Is indentation necessary for Python program?

One of the most unique features of Python is its indentation. In Python, a block of code can be uniquely identified by its indentation. Unlike C or Java language, the block of code is identified using curly braces. In such case, the indentation of the code is not mandatory. Due to this and sometimes due to bad programming practice, it is quite difficult to read the program block in C and Java. Python makes it mandatory, and thus, it improves the readability of the program code.

10. What is the recommended pip to install Python packages?

It depends upon the Python version. The pip is preinstalled for Python 2.7. But if we are using Python 3 version, it is recommended to use pip3. Before installation, we can check the version of the same by the command pip version. If it is not installed, we should bootstrap from the standard library using the command `python -m ensurepip-default -pip`. Then, install it as per the requirement.

11. In Linux, do Python packages be installed through Synaptic Package Manager?

Synaptic Package Manager is good to install the standard Python package. But its major disadvantage is that it is dependent upon the versions. In the low version of synaptic, it is quite difficult to get the latest version of the package. The same problem may appear for the Ubuntu software center. The best approach for the latest Python package is to install it via the pip installer.

12. In case of machine learning code development, is it necessary to use NumPy and SciPy package?

The elementary machine learning methodology is used for more complex mathematical computation. NumPy gives us a good amount of numerical computation as well as data structure definition. SciPy offers more complex mathematical computations such as integration and differentiation. Besides, pandas and scikit-learn are the essential ingredients of fundamental machine learning.

13. What is the difference between command-line arguments of Java and Python?

   The major difference is the approach of the command-line argument. In case of Java, command line argument is to be passed through the `public static void main(String args[])` method. In this case, `String args[]` is the command-line argument of string type. To use the command-line argument in Python, we have to import the sys package. Then the argument can be found from the `argv` list of the `sys` package. Here, in this case, we should write `x = sys. argv[0]`.

14. Is it possible to run two Arduinos in parallel mode?

   Two Arduinos can be coupled in master salve mode. To do that, we can make an inter integrated circuits (I^{2}C) communication interface between the two Arduinos. This is mostly done if we want to share the workload of a single Arduino. In this case, serial data (SDA) and serial clock (SCL) have to be connected for both Arduinos. Nowadays, multiple codes can be run parallel to an Arduino same time by virtue of the multitasking real time operating systems (RTOS) for Arduino support. A free multitasking RTOS is now available so that you can run the core in parallel.

15. Is it possible to run Tensor Flow or another deep learning tool in Raspberry Pi?

   Deep learning tools such as like Tensor Flow, PyTorch, and Keras deal with highly complex convolution neural network (CNN) and recurrent neural network. These models take an enormous space for computation. For example, aneural network uses 200 MB for the floating-point format sometimes due to the weight of the neural connection. There are millions of such connections in a network. The weight perhaps is stored in a 32-bit floating-point format. We can perform quantization to shrink the file size and make a min and max for each layer. In this way, the compression of the weight value may be possible, which is perhaps suitable for running a complex CNN in a machine that has low memory capacity.

16. Is there any separate hardware available for Tensor Flow and deep learning?

   There are several dedicated hardware items available for Tensor flow machine learning. Popular among them is Lambda Deep Learning Dev Box consisting of Ubuntu 18.04 bionic OS, 4xNVIDIA Gforce GTX 1080 Ti, Intel XEON processor with 64 GB DDR with 2GB SATA SSD, and 4 GB of HDD. It has preinstalled Keras, Theano,

Tensor Flow, and Pytorch. Google Edge TPU is another invention for high-power machine learning work for a low-power computing device like Raspberry Pi. With the help of such TPU, we can create lightweight AI platform in the portable device level. Its CPU is NXP i.MX 8M SoC (a quad Cortex-A53, M4F). Its GPU is integrated GC7000 Lite. It comprises 1GB RAM and 8GB flash. The supported OS is Debian Linux and Android Things. The Tensor Flow Lite framework is also supportable. Edge TPU Accelerator is another platform that can enhance the Tensor Flow processing capability of the application by introducing it as a USB device. The specification is the same as the previous one.

Index

Printed and bound by CPI Group (UK) Ltd, Croydon, CR0 4YY

17/10/2024

01775661-0001